Vision and Attention

Springer
New York
Berlin
Heidelberg
Barcelona
Hong Kong
London
Milan
Paris
Singapore
Tokyo

Michael Jenkin
Laurence Harris
Editors

Vision and Attention

With 92 Illustrations

Springer

Michael Jenkin
Department of Computer Science
York University
4700 Keele Street
Toronto, Ontario M3J 1P3
Canada
jenkin@cs.yorku.ca

Laurence Harris
Department of Psychology
York University
4700 Keele Street
Toronto, Ontario M3J 1P3
Canada
harris@yorku.ca

Cover illustration: The cover art shows frames taken from the film *Gorilla Thump*, which can be found on the CD.
Cover and CD illustration © 1999, Daniel J. Simons. Used with permission.

Library of Congress Cataloging-in-Publication Data
Vision and attention / edited by Michael Jenkin, Laurence Harris.
    p. cm.
   Includes bibliographical references and index.
   ISBN 0-387-95058-3 (alk. paper)
    1. Attention—Congresses.   2. Visual perception—Congresses.   I. Jenkin, Michael
(Michael Richard McLean), 1959–   II. Harris, Laurence (Laurence Roy), 1953–
QP405 .V574 2000
155.14—dc21   00-059474

Printed on acid-free paper.

Production managed by Jenny Wolkowicki; manufacturing supervised by Erica Bresler.
Photocomposed pages prepared from the editors' files.
Printed and bound by Maple-Vail Book Manufacturing Group, York, PA.
Printed in the United States of America.

9 8 7 6 5 4 3 2 1

ISBN 0-387-95058-3   SPIN 10770128

Springer-Verlag   New York Berlin Heidelberg
A member of BertelsmannSpringer Science+Business Media GmbH

To HLM and CMO.

Preface

This book is based on a conference on Vision and Attention, the fifth conference of the York Centre for Vision Research organized by I. P. Howard, D. M. Regan and B. J. Rogers in June 1999 and sponsored by the Centre for Vision Research, the Departments of Psychology and Computer Science, of York University.

The CD-ROM that accompanies this book contains colour imagery and video clips associated with various chapters and the conference itself. The CD-ROM is presented in HTML format and is viewable with any standard browser (eg. Netscape Navigator or Microsoft's Internet Explorer). To view the videos on the CD, you will need Quicktime, which is available free from Apple. To view the CD, point your browser at the file **index.htm** on the CD. Please note that some of the larger movies require a machine with 128Mbytes of memory in order to play them correctly.

The cover art shows frames taken from the film *Gorilla Thump*, which can be found on the CD and which is Copyright 1999, Daniel J. Simons. Used with permission.

The York Vision Conference, and this book, would not have been possible without the advice and support of Ian P. Howard, David Martin Regan and the Human Performance in an Aerospace Environment Theme of the Centre for Research in Earth and Space Technology (CRESTech). Behind any successful endeavour is the person who really runs things, and none of this would have been possible without Teresa Manini.

York University   Michael Jenkin
October 2000   Laurence Harris

Contents

7   Motion-Disparity Interaction and the Scaling of Stereoscopic Disparity   129

Michael S. Landy and Eli Brenner

8   Signal Detection and Attention in Systems Governed By Multiplicative Noise   151

Christopher W. Tyler

Contributors

Narcisse P. Bichot, Vanderbilt Vision Research Center, Department of Psychology, Vanderbilt University, Nashville, TN 37240, USA.
narcisse.p.bichot@vanderbilt.edu

Eli Brenner, Vakgroep Fysiologie, Erasmus Universiteit, Postbus 1738, 3000 DR Rotterdam, The Netherlands.
brenner@fys.fgg.eur.nl

Heinrich H. Bülthoff, Max-Planck-Institute for Biological Cybernetics, Spemannstraße 38, 72076 Tübingen, Germany.
heinrich.buelthoff.@tuebingen.mpg.de

John M. Findlay, Centre for Vision and Visual Cognition, Department of Psychology, University of Durham, South Road, Durham, DH1 3LE, England.
j.m.findlay@durham.ac.uk

Iain D. Gilchrist, Department of Experimental Psychology, University of Bristol, 8 Woodland Road, Bristol, BS8 1TN, England.
I.D.Gilchrist@bristol.ac.uk

Laurence R. Harris, Department of Psychology, York University, 4700 Keele Street, Torotno, Ontario, M3J 1P3, Canada.
harris@yorku.ca

Michael J. Hawken, Center for Neural Science, New York University, 4 Washington Place, New York, NY 10003.
mjh@cns.nyu.edu

Michael Jenkin, Department of Computer Science, York University, 4700 Keele Street, Toronto, Ontario, M3J 1P3, Canada.
jenkin@cs.yorku.ca

Elizabeth N. Johnson, Center for Neural Science, New York University, 4 Washington Place, New York, NY 10003, USA.
zab@cns.nyu.edu

Radha P. Kohly, Department of Biology, York University, BSB, Room 375, 4700 Keele Street, Toronto, Ontario, M3J 1P3, Canada.
radha@yorku.ca

Michael S. Landy, Department of Psychology and Center for Neural Science, New York University, 6 Washington Place, New York, NY 10003, USA.
landy@nyu.edu

Sally McFadden, Department of Psychology, University of Newcastle, Newcastle, NSW, Australia.
mcfadden@psychology.newcastle.edu.au

Stephen R. Mitroff, Department of Psychology, Harvard University, 33 Kirkland Street, #820, Cambridge, MA, 02138, USA.
mitroff@wjh.harvard.edu

J. Kevin O'Regan, Laboratoire de Psychologie Expérimentale, CNRS, EHESS, EPHE, Université; René Descartes, 71, avenue Edouard Vaillant, 92774 Boulogne-Billancourt Cedex, France.
oregan@ext.jussieu.fr

David Regan, Department of Biology and Psychology, York University, 4700 Keele Street, Toronto, Ontario, M3J 1P3, Canada.
dregan@yorku.ca

Ronald A. Rensink, Cambridge Basic Research, Nissan Research & Development, Inc., 4 Cambridge Center, Cambridge, MA 02142-1494, USA.
rensink@pathfinder.cbr.com

Dario L. Ringach, Center for Neural Science, New York University, 4 Washington Place, New York, NY 10003, USA.
dario@ucla.edu

Michael P. Sceniak, Center for Neural Science, New York University, 4 Washington Place, New York, NY 10003, USA.
sceniak@cns.nyu.edu

Robert M. Shapley, Center for Neural Science, New York University, 4 Washington Place, New York, NY 10003, USA.
shapley@cns.nyu.edu

Daniel J. Simons, Department of Psychology, Harvard University, 33 Kirkland Street, #820, Cambridge, MA, 02138, USA.
dsimons@wjh.harvard.edu

John K. Tsotsos, Department of Computer Science and Centre for Vision Research, York University, 4700 Keele Street, Toronto, Ontario, M3J 1P3, Canada. tsotsos@cs.yorku.ca

Christopher W. Tyler, Smith-Kettlewell Eye Research Institute, at California Pacific Medical Center, 2318 Fillmore Street, San Francisco, CA, 94115, USA. cwt@skivs.ski.org

Hendrik A. H. C. van Veen, TNO Human Factors Research Institute, P.O. Box 23, 3769 ZG Soesterberg, The Netherlands. vanVeen@tm.tno.nl

Josh Wallman, Department of Biology, City College, City University of New York, New York, NY, USA. wallman@sci.ccny.cuny.edu

Frances Wilkinson, Department of Psychology, Atkinson College, York University, 4700 Keele Street, Toronto, Ontairo, Canada, M3J 1P3. franw@yorku.ca

Hugh R. Wilson, Department of Biology, York University, 4700 Keele Street, Toronto, Ontario, Canada, M3J 1P3. hrwilson@yorku.ca

1

Vision and Attention

Laurence R. Harris
Michael Jenkin

The term "visual attention" embraces many aspects of vision. It refers to processes that find, pull out and may possibly even help to define, features in the visual environment. All these processes take the form of interactions between the observer and the environment: attention is drawn by some aspects of the visual scene but the observer is critical in defining which aspects are selected.

Although this book is entitled *Visual Attention*, none of the processes of "visual" attention are exclusively visual: they are neither driven exclusively by visual inputs nor do they operate exclusively on retinal information. In this introductory chapter, we outline some of the problems of coming to grips with the ephemeral concept of "visual attention."

1.1   What Is Attention?

Attention implies allocating resources, perceptual or cognitive, to some things at the expense of not allocating them to something else. This definition implies a limit to the resources of an individual such that they cannot attend to everything at once, all the time. In one sense this is obvious in that the senses already provide a filter. The visual system, for example, does not tell us about what is happening behind us or in the infrared part of the spectrum. But attention refers to selection from the array of information that is arriving at the brain and is potentially available.

The term "attention" suffers from the fact that it is a word in both common and scientific usage and the common and scientific meanings only loosely overlap. Furthermore, both common-parlance and scientific "attention" each have several different meanings. The common usage has an implication of urgency and alertness well - illustrated by the single-word sentence substitution "ATTENTION!!." In scientific usage, we do not necessarily want to incorporate this sense of the unusual, although we cannot avoid the tint.

Although closely related to "active vision" (see Aloimonos et al., 1988, and also Harris and Jenkin, 1998), attention describes the requirement that the visual system attends at least at a computational level (although not necessarily a physical level) to different events in the visual field. Active systems may require attentional mechanisms in order to direct their sensors at different salient events, but attentive systems are not necessarily active in a physical way.

The word "attention" has ancient roots that link it with concepts of general alertness and conscious receptivity on the one hand and with concentration and focusing on the other. Like other words imported from common parlance, for example "stress," "attention" can perhaps only be meaningfully considered if accompanied by an adjective. Just as there is no such thing as unqualified "stress" physiologically, it might be useful to start from the stance that there is no such thing as unqualified attention. What types of "attention" are there? We suggest four distinct types:

- **Selective attention**. At any one time we seem to attend to and be aware of only some aspects of the sensory input. What is selected is sometimes determined by the demands of a particular task, such as concentrating on a tool while it is being employed. But selective attention is not only activated by interest in a goal. Although it is obvious that we are selectively attending while actively searching for those lost keys, selection is a feature central to the act of seeing anything. It is impossible, with our meagre brain equipment, to process the whole of the retinal image. Selectively attending to something implies that the feature being attended to has already been defined, whether this is a basic attribute such as a colour, a visual direction, or an actual object.

- **Parsing attention**. Attention might be a part of the process of recognizing objects and separating them from their backgrounds. It has been suggested that the act of attending to something is critical to the binding together of the the various features that define a perceptual object.

- **Directing attention**. When something happens, a primitive reflex system instinctively orients us towards it. This is an emergency, interrupting system, sometimes called attentional capture, which overrides normal behaviour when something potentially important or dangerous demands immediate analysis. But directing attention is also a more gentle, omnipresent drive used for exploring the environment or for maintaining attention on an object while carrying out a task. There might be a continuum between emergency interrupt, normal exploration, and maintaining attention or perhaps these behaviours represent quite different control systems.

- **Alertness attention**. It seems intuitive that a certain level of arousal is necessary for normal perceptual processes to operate. Perception is normally associated with a state of being awake and responsive to sensory input: a state in which behaviours are planned and carried out and in which we are interacting with the environment. But how much undirected arousal is in fact necessary for perception? Some perception can occur while in a daydream or even while asleep (Mack and Rock, 1998). Alertness attention may operate by modulating other forms of attention.

1.1.1   Should "attention" be regarded as a discrete behaviour?

Should attention be regarded as a discrete behaviour that can occur independently of other behaviours? Can the act of attending to something be treated analogously to the discrete act of picking something up between the thumb and forefinger or directing the eyes to converge on a given point? It is possible for finger control or auditory localization to occur quite independently of any other behaviour. Can "attending" be regarded similarly or is it unable to exist independently of other behaviours such that it cannot be regarded as having an independent existence? Perhaps attention is just a modulation of other behaviours.

This is a significant question when seeking to understand attention because if attention can be regarded as a discrete behaviour, then it might be expected to have special brain mechanisms and perhaps brain sites, devoted to it. If, on the other hand, it exists only as a modulation of other behaviours, then no such specialized sites would be expected. Should we seek control systems in the brain that specifically underlie the allocation of attention in the same way as we expect to find systems devoted to auditory localization or locomotion control?

This debate often takes place in the context of whether there is an executive controller. The controller would not be directly involved in any sensory processing but only in the control or modulation of such target systems when it applied attention. Many parts of the visual sensory system show clear modulation in response to attention (e.g., parietal cortex, superior colliculus, even lateral geniculate and striate cortex - see chapter 11 by Bischot), but these areas also show stimulus-related activity when attention is not involved. One area, the cingulate cortex however seems to show activity exclusively when attention is required. Hence, it is a strong candidate for the location of an executive controller (Badgaiyan and Posner, 1998; Carter et al., 1998). When a task that previously required attention becomes automatic through practice, activation of the cingulate cortex is no longer found (Frith et al., 1991).

This therefore gives us a model for what attention is and how it is implemented by the brain. The model is driven by a central executive controller, located at least partially in the ancient limbic association areas, that either makes or at least administers decisions. The decision about what is to be attended is then implemented by selective modulation of the activity or sensitivity of the sensory area or areas that process the desired attribute. The modulation is as specific or fine-tuned as required for the task that the controller is executing. The resulting increased activity in the processing system then gives that coded attribute a competitive advantage or even a flag that leads to preferential treatment. Why might selected attributes need preferential treatment?

1.2   Selective Visual Attention

The amount of information falling on the retina at any one time is truly vast and most of it is of no survival value at all (Fig. 1.1). Look at the heavens and all the

FIGURE 1.1. Bookshop scene. The amount of information falling on the retina at any one time is enormous. A selective process is required not only to acquire useful information but in order to be able to see at all. Often the selection will depend on the task at hand. A person just looking at this photo is unlikely to be able to report on the type of lighting, for example, unless attention was specifically drawn to that type of information.

visible stars within the eyes' visual fields are imaged on the retina at once. Look at a tree with its complex and detailed branching pattern, much of which can be resolved at the same time. Look at this page where all the letters are focused on the retina. Most of the potentially smothering information arriving at the retina is lost by the limited resolving power of the retina outside the central few degrees of the fovea. But even with this blesséd filter, there remains an enormous amount of information that could, theoretically, be extracted. Normally, the gaze only stays at one point for about 300 msecs (Yarbus, 1967), implying that the information needed is extracted in this small time before the gaze is shifted again.

Visual attention is an essential component of machine-vision systems. Figure 1.2 shows the Eyes 'n Ears sensor (Kapralos et al., 2000). This is an omnidirectional sensor based on the Paracamera. Consisting of both audio and video components, the Eyes 'n Ears sensor's visual input is generated by a video camera mounted vertically and directed at a semispherical mirror. The resulting view (shown in Fig. 1.2b), provides a 360° view of the environment. This view can be unwarped to provide a perspective panoramic view (Fig. 1.2c). As in the biological example of Figure 1.1, it is not practical for a machine to attend to each and every pixel in the view. Many are uninteresting for the specific task at hand. For the Eyes 'n Ears sensor, locations in which the image changes are of interest, so the sensor uses an attention-directing mechanism based on attending at image locations that correspond to image differences. Figure 1.3 shows the system in operation. Figure 1.3c is a retinotopic difference image between the image frames shown in Figs. 1.3a 1.3c. Brighter locations correspond to more interesting image locations.

At any one time, the brain just does not need to know very much about the world - the details of the branching structure of a tree are usually of no use and therefore

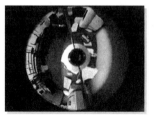

(a) Sensor   (b) Camera view

(c) Perspective (unwarped) image

FIGURE 1.2. Eyes 'n Ears sensor.

of no interest. At first glance, it might seem useful to take in the entire content of a page of text at once, like Commander Data of *Star Trek*, but understanding that information is a serial task requiring breaking down, or parsing, the retinal pattern into words and interpreting each word in its correct sequence. This requires selecting the important features (features such as cross-strokes and uprights that define letters, for example) and discarding others (such as the colour or details of the font or the particular layout of the words on the page or paper).

A selective mechanism saves the brain valuable time and processing capacity, allowing it to dedicate its limited resources toward doing something with survival value. It also gives certain information a "hotline" so that it is processed preferentially, resulting in faster and more accurate reactions (Posner et al., 1978) and better sensitivity (Carrasco et al., 2000 – see Chapter 3 by Hawken et al. in this volume). This is not just a general alertness, but a specific visual process because only objects that appear at the expected visual location enjoy enhanced reaction times (Posner, 1980).

At any one time, some information in the sensory array is important and some is not. What is and is not important are by no means simple for the experimenter, or a naturally behaving brain, to determine. Furthermore, what is significant will vary from species to species, from individual to individual, and from moment to moment, even for a given individual and a given scene. If survival is benefitted by filtering salient information from a potentially overwhelming sensory bombardment, what exactly is it that is extracted? How is that done? And what is the connection between "selection" and "attention"?

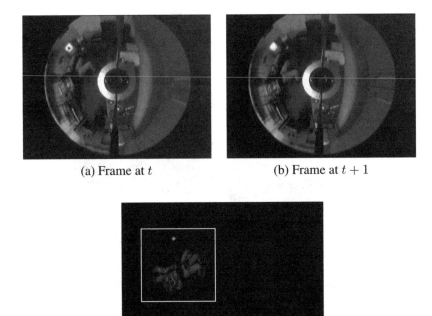

(a) Frame at t (b) Frame at $t + 1$

(c) Difference image

FIGURE 1.3. Attending with the Eyes 'n Ears sensor.

1.2.1 What is selected?

People and animals can choose to attend to different things. They can attend to a modality (vision, hearing, touch, taste, smell) or to a colour, or shape or location. Intuitively it seems that the act of attending to one thing makes other things less noticeable - even an entire modality can be ignored in this way.

Outside the laboratory, a specific decision to concentrate on a single attribute is not usually required - the demand arises naturally from the task at hand. When trying to thread a needle, less attention is paid to the radio or the colour of the thread than to the lining up of the thread with the hole in the needle. This does not mean that other things are not processed at all. If something significant happens in an unattended modality (such as hearing one's name), it will not go unnoticed (Moray, 1959; McCormick, 1997). This means that unattended things must be processed to a high enough level to identify them but with a disadvantage or attenuation that normally keeps them unnoticed (Treisman, 1960).

When attending, what actually is being attending to? Is attention directed to a particular zone of space in which interesting things are occurring? This is the implication of the spotlight (Posner, 1980) or zoom-lens (Eriksen and St. James, 1986) models of attention. Under these models, anything appearing in the attended zone enjoys the advantages of attention. Alternatively, perhaps a particular object

(a)   (b)

FIGURE 1.4. Berkeley can attend either to (a) a location, in this case a hole, or (b) an object, in this case an attractive doggie treat.

or feature is selected? This would be appropriate for a task-based role for attention. In this case, even if two objects appear close together, only the attended one will receive the full benefits of selection. These viewpoints are reviewed in Chapter 2 by McFadden and Wallman and their differential predictions are illustrated in Fig. 1.5. If attention is object based, then points on the same object (e.g., Point b in Fig. 1.5) should receive more benefits than points on other objects (e.g., Point c) once any part of that object has been selected (e.g., Point a). On the other hand, a spotlight-like, spatially defined attentional beam would radiate symmetrically from its focus, indiscriminately selecting whatever is within the zone and thus giving equal advantage to points b and c (Fig. 1.5).

Curiously enough, the answer does not seem to be straightforward. Whether attention is location- or object-based probably depends on what the subject is trying to do. While attention can be object based (Duncan, 1984; Egly et al., 1994; Weber et al. 1997; Moore et al., 1998), sometimes, as when Laurence's dog, Berkeley is hoping something will come out of a hole (Fig. 1.4), or when a subject is cued to attend to a particular part of a CRT screen (Posner, 1980), it seems possible to attend to an empty location (Egeth and Dagenbach, 1994). On the other hand, when Berkeley attends to a hole, it is possible that other stimuli are attracting or guiding his attention rather than just a volume of space. The visual cues that define the hole are clearly relevant, although they are not the cues attended to as such - they are merely guiding the attention such that a speedy response to the emergence of something interesting can be guaranteed. For the subject viewing the CRT screen, there are also visual cues specifying the location and distance of the attended point on the screen.

We can conclude that visual items can be selected, although the details of what exactly is being selected are still not settled. But how does either a either particular object or a location come to be selected from all its competitors?

1.2.2   How is selection achieved? How much salience is due to the sensory input itself and how much to higher processes?

Of course there is nothing salient about the environment at all. Salience, like colour and beauty, is in the eye of the beholder. Even the most dramatic and attention-grabbing visual events seen on a TV screen pass seemingly unnoticed by Laurence's dog, Berkeley, and fire hydrants, although of great interest to Berkeley, scarcely attract human attention at all. Salience implies some kind of "biological significance".

How is such biological significance to be defined and determined? There is probably only one general rule for stimulus-determined biological significance: "Has something changed?" Although change is neither necessary nor sufficient to make something interesting it clearly indicates something that might not have been previously explored or that might have moved and therefore altered its status as something of interest. One way of assessing the significance of features is simply to present stimuli and see what grabs human attention!

When we look at an image, certain patterns, shapes, or areas defined by various parameters seem to pop out instantly and effortlessly, without the need for searching (Fig. 1.6). In a simple pattern, such as the characters in Figs. 1.6a and 1.6b, this feature cannot be missed or ignored. In a more complex scene, such as a crowded bookshop scene (Fig. 1.1), what is immediately apparent is not any particular pattern but a gist: that it is a bookshop. Neither of these perceptions require careful examination of the scene but instead seem to indicate a global, pre-attentive, parallel processing of the visual information (see Chapter 4 by Wilkinson and Wilson) all at once applied to the entire scene before detailed conscious attention, of the type discussed in the previous section, is deployed (Treisman and Gelade, 1985). These are examples of what perception can do without attention. They are what the next stage, attention proper, must work on. When attention is applied, presumably the popped out features are the most likely to be selected. That would imply that the route of attention is determined by the stimulus.

The question of the relative significance of intrinsic stimulus-based features as opposed to cognitively imposed salience has been explored experimentally by artificially attaching a rather tepid salience to inert little pictures – not biologically but by request. Subjects were asked for example, to search for inverted T's hidden in a mass of distractor items (e.g., Eriksen and Hoffman, 1972, 1973; Posner, 1980). In such tasks, the intrinsic salience is in the feature content that distinguishes the target from its surroundings (Treisman and Gelade, 1988), and the cognitively added salience is that it had been consciously selected as a target.

Change is probably the most reliable, stimulus-based attention attractor, especially when it indicates a new object (Yantis and Jonides, 1984; Yantis and Hillstrom, 1994; Yantis, 1998). Most other cues (Jonides and Yantis, 1988; Theeuwes, 1990), and, surprisingly, even object motion (Hillstrom and Yantis, 1994), turn out not to be reliable stimulus-based attractors. The conclusion is that when looking around a fairly homogeneous scene or photograph, there is usually very little in the scene itself to which everyone would have their attention drawn.

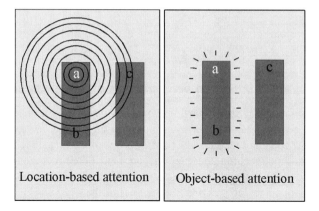

FIGURE 1.5. Spread of attention around a location (left) or around an object (right). The points b and c are equidistant from a, but a and b are on the same object, whereas a and c are on different objects. If attention is object-based, then attending to point a should be beneficial to point b because it is part of the same object. If there is just an indiscriminate spatial spread of attention around location a however, the points b and c should benefit equally.

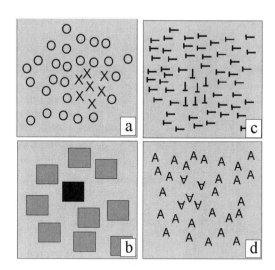

FIGURE 1.6. Examples of pop-out stimuli. Areas within each figure can be detected with different degrees of ease. The areas here are defined by shape (a), contrast and luminance (b), or orientation (c and d). The areas defined are visible immediately (pop out) in some cases (a and b). Not all orientation changes are sufficient to define a form, however, as illustrated in (d).

Salience is thus a varying property of features in the environment determined by an interaction between them and the observer. Salience is determined neither wholly by the properties of the object nor by the state of the observer but by the interaction between them. A food item is only a food item if you are hungry.

1.2.3 What is the connection between selection and attention?

The retinal input is used by many quasi-independent systems for a variety of tasks that are carried out simultaneously. Visual information is used, for example, for pupil control, directing and maintaining fixation, controlling head and body orientation, and for guiding reaching, locomotion, and many other interactions with the environment. Although research in the last few years has elevated these behaviours above the status of mere reflexes and shown them to be quite sophisticated and responsive to different contexts, nevertheless they largely take place without our knowledge or attention. Each of these tasks has particular needs, which involve extracting different features from the retinal input and elsewhere. Mostly, the systems need only to know some special pieces of information and those only very transiently. The fact that your hand is presently three inches from the coffee cup with a time-to-contact of 0.7 sec. is just not something that you need to know at any time other than the present. It is not necessary to store information of this type beyond the time that it is useful (see Chapter 4 by O'Regan). In addition to these various, essentially subconscious, visual functions, however, the existence and details of some visual objects reach perceptual awareness.

Although all tasks in which a person or animal is involved require some selection, only some need and receive attention. Attention implies awareness. Selective attention thus refers only to those selection processes that present their selections to awareness. Why should we become aware of anything? What could be the biological survival value of awareness? Why might some filtered aspects of the sensory information pass into conscious awareness while most of what is not rejected out of hand is dealt with transiently by autonomous mechanisms? The essence is that awareness implies continuity. It is only possible to have an awareness of the present in the context of a past and an anticipation of a future. Awareness then becomes intimately associated with memory (see Dennett, 1991). Memory links the awareness of the present moment to past experience in a way that is necessary for true, context-laden awareness to occur. Further, awareness of the present moment viewed in the context of previous experience allows prediction of the future. Awareness can thus be seen as an emergent property of a selective memory process. Things that are selected then are able to interact with previous memories and might themselves be laid down as memories.

This argument does not outline an exclusive connection between attention, memory, and awareness. It is not necessary that everything must be selected and go through awareness to be remembered or for predictions to be made.

Awareness and memory allow us a degree of flexibility that is difficult to incorporate into an algorithm. It enables us, for example, to acknowledge but to not respond to the repeat of an event that was previously deemed interesting. Awareness allows the application of history and an anticipation of the future. Interest and importance can be added to and removed from the environment at will: sticks can become interesting and scarecrows can be ignored. By this argument, attention becomes the tool of a selective awareness: a high-level process that guides perception. Research into attention thus becomes an exploration of the limits and implementation of awareness. An excellent review of many of these issues can be found in Palmer (1999).

1.2.4 Mechanisms of selective attention

Selective attention is controlled by mechanisms on at least four levels:

1. The mechanism that leads a high-level executive controller to decide to attend to (or to ignore) something. This may be a location, a class of objects, a modality, or a single object, depending on the behaviour in which the subject is engaged.

2. The mechanism by which neural activity related to the target items is given a selective advantage.

3. The mechanism that detects the appropriately enhanced or otherwise flagged activity and passes it to awareness.

4. How awareness uses that information.

Imaging techniques have suggested brain sites for parts of mechanisms 1 and 4, although the details of how they might work are unknown. Most research has concentrated on mechanisms 2 and 3 with exciting results that allow us to relate neural behaviour to perception.

Visual sensory neurones are tuned to particular properties of the stimuli to which they respond. Often, these cells show a spatial tuning, which takes the form of a receptive field: an area of the retina outside which stimuli will evoke little or no response (see Chapter 3). They may in addition be tuned to orientation, wavelength, direction and speed of motion, spatial frequency, or binocularity - all the parameters that describe the image. Different cells in different areas are selective for different attributes (Zeki, 1978). Some cells are tuned for complex combinations of features that provide a high-level description of the scene, such as faces in particular orientations (Perrett et al., 1985). In response to the presentation of a stimulus with a particular combination of features, a given cell will have a particular firing rate. The presence or absence of a cell's firing above the background level is an indicator that those features that it is tuned for, are present (Barlow, 1972; Rieke et al., 1997).

Mechanism 2, the mechanism by which neural activity is given a selective advantage, seems to involve an enhancement of a cell's normal response. For a given stimulus configuration, a larger response than usual is evoked when attention is being paid to those features coded by that cell. Visually responsive cells in many areas show an enhancement if a stimulus that normally excites them is attended (Motter, 1998; Colby and Goldberg, 1999; see Bischot's Chapter of this volume). The enhancement is the signature of the neuronal workings of attention, presumably making the activity easier to detect by the next stage of processing (see Chapter 8 by Tyler). Desimone and Duncan (1995) suggest a competitive neural network for extracting this enhanced activity. Their model has recently received considerable support and is further explored by Wilkinson and Wilson in Chapter 4. A more general model is given by Tsotsos in Chapter 6. One of the problems the enhancement-detection mechanism must face is how to spot an enhanced response in the presence of other responses evoked by other stimulti that, although not attended, might still be powerfully activating their own detecting systems. For example, when searching for a blue object, imagine the brain state when a small, only slightly blue object is spotted among many other brightly coloured, high contrast items that, even unenhanced, would be activating many neurones strongly. How is the weak response enhanced enough to be chosen?

Notice that various other selection procedures referred to in the previous section operate simultaneously and in parallel with the selective attention process outlined here. These other selection processes might share some of the mechanisms of stage 2 and lead to enhancements of their own. Neural activity for targets that are going to be used for the next saccade will be enhanced (Wurtz and Mohler, 1976a, b) but may not necessarily be selectively attended.

1.3   Parsing Attention. Is Attention Central to the Act of Seeing or is it Merely a Servant Carrying its Master to the Right Place?

We have argued that selective attention is employed in the act of bringing things to conscious awareness. Is the act of attention actually a necessary part of defining perceived objects (Yantis, 1998; Regan, 2000)? The various features of the visual world are carried by a whole population of different cells, each involved in coding a different feature of the image. Starting from the retina itself where each cell responds only to features appearing within its discrete receptive field, the processing of the image is spread out over many cells. At higher levels, cells become more specialized in the attributes that they are involved in processing (Zeki, 1978). Different systems encode colour, motion, depth, etc. How this scattered information is put back together to produce a coherent percept of each individual object is called the binding problem. Parsing attention might be an integral part of the solution to the binding problem. Under this theory, the perceptual world is not parsed into individual objects until the act of attending to them somehow binds their features and separates them from their background (Treisman, 1998). This possibility is developed by Regan and Kohly in Chapter 13.

If attention is needed to bind features together to create perceptual objects, then this produces a logical problem for those theories in which selective attention is attracted not to locations but to objects (Duncan, 1984; Egly et al., 1994; Weber et al. 1997; Moore et al., 1998). It would seem under these theories, that objects need to be defined before they can attract attention, but that attention is required to define the object. This conundrum can be resolved by having multiple types of attention as we postulate in this chapter. Directing attention (see section 1.4) works with primitive object definition using only the visual analysis available with the poor resolving power of the peripheral retina. Subsequent application of parsing-attention can aid the object-definition process.

One way to help understand the role of attention in vision is to explore what, if anything, can be perceived without it (see Mack and Rock, 1998). Recently the theme of "change blindness" has highlighted the need to attend consciously to something if it is to be seen at all. If, during some minor distraction or break in attention, a scene is modified, even quite dramatically, that change is often difficult or impossible to spot without many repetitions of the change (see Chapters 9, 10 and 14). Change blindness has been interpreted as illustrating the need for directed attention in order for perception to occur. A representation of the first scene must be retained (memorized) to compare with the second. Change blindness demonstrates how little of a scene is actually memorized normally. But the change blindness paradigm is unnatural in two ways: First, normal vision must be disrupted (by a flash, mud splash, or brief temporal interval) and secondly attention guidance (mechanism 1 in section 1.2.4) is severely hampered because the target for attention is not specified.

1.4   Directing Attention

The metaphor of attention as a spotlight scanning around the visual world (Posner, 1980) naturally links it to the act of directing the gaze. Although the spotlight metaphor does not explain all aspects of selective attention (see Driver and Baylis, 1989 and Chapter 13, for example) it is an intuitively attractive idea and has been very influential in studying attention. It seems that attention does move as a spotlight at least sometimes (Tsal, 1983; Posner and Peterson, 1990; see Chapters 2 and 5 of this volume). Although attention movement is normally linked to gaze control, the spotlight can be moved covertly, independently of actual gaze movement.

Clearly, to be useful, visual attention must be directed. But does the mechanism that does this qualify as a distinct type of visual attention in itself or is it just a transport mechanism? The act of orienting does not need to be powered by vision. Events involving touch or sound often demand orientation even more urgently than vision, and orienting movements do not have to be exclusively visual. An itch on the neck can only be reached with the hand or foot (in Laurence's dog Berkeley's case) and is never available to direct visual scrutiny. This suggests an orienting system independent of the awareness-defined selective visual-attention system and its concern with objects.

Looking around a static scene uses directing attention to select different parts of the scene. These mild, nonurgent orienting movements of the gaze will be driven by a mixture of internal and external stimuli. Internal drives might include a task, such as reading or long-term plans. Attention might be drawn, for example, to a particular book that was of interest weeks earlier but that was spotted while gazing around aimlessly. External stimuli include dramatic, attractive things such as movement or flashes of light (as well as sounds, smells, and tickles) but also features such as edges or corners. This interaction between internal and external contributors to what drives the orientation system is similar to the discussion of salience in the visual-attention system (Section 1.2.2). Are they the same? This question is addressed by McFadden and Wallman in Chapter 2. They conclude, on the basis of adaptation studies, that they are closely linked, whereas Findlay and Gilchrist (Chapter 5) suggest they are more independent.

1.5   Conclusions

The enormous interest in the topic of visual attention reflects the willingness of researchers to consider vision in real life (see Chapter 12 by Bülthoff and van Veen). Rather than treating people as psychophysical machines who always have the same responses to a given stimulus, scientists now freely acknowledge the role of attention and the predominant role of mind-set and higher-level cognitive factors in even very simple visual tasks. How attention contributes to and interacts with perception takes us close to some of the big questions of the nature of consciousness and the connection between neural activity and perception: mind and brain.

References

Aloimonos, J., Weiss, I. and Bandopadhyay, A. (1988). Active vision. *Int. J. Comp. Vis.*, 1:333-356.

Badgaiyan, R. D. and Posner, M. I. (1998). Mapping cingulate cortex in response to selection and monitoring. *Neuroimage*, 7:255-260.

Barlow, H. B. (1972). Single units and perception: a neuone doctrine for perception. *Perception*, 1:371-394.

Carrasco, M., Penpeci-Talgar, C. and Eckstein, M. (2000). Spatial covert attention increases contrast sensitivity across the CSF: support for signal enhancement. *Vis. Res.*, 40:1203-1216.

Carter, C. S., Braver, T. S., Barch, D. M., Botvinik, M. M., Noll, D. and Cohen, J. D. (1998). Anterior cingulate cortex, error detection and online monitoring of performance. *Science*, 280:747-749.

Colby, C. L. and Goldberg, M. E. (1999). Space and Attention in Parietal Cortex. *Ann. Rev. Neurosci.*, 22:319-349.

Dennett, D. (1991). *Consciousness Explained*. Boston: Little Brown.

Desimone, R. and Duncan, J. (1995). Neural mechanisms of selective visual attention. *Ann. Rev. Neurosci.*, 18: 193-222.

Driver, J. S. and Baylis, G. C. (1989). Movement and visual attention: the spotlight metaphor breaks down. *J. Exp. Psychol.: Hum. Percept. Perform.*, 15:448-456.

Duncan, J. (1984). Selective attention and the organization of visual information. *J. Exp. Psychol. Gen.*, 113:501-517.

Egeth, H. E. and Dagenbach, D. (1994). Shifting visual attention between objects and locations: evidence from normal and parietal lesion subjects. *J. Exp. Psychol. Gen.*, 123:161-177.

Egly, R., Driver, J. and Rafal, R. D. (1994). Shifting visual attention between objects and locations: evidence from normal and parietal lesion subjects. *J. Exp. Psychol. Gen.*, 123:161-177.

Eriksen, C. W. and Hoffman, J. E. (1972). Temporal and spatial characteristics of selective encoding from visual displays. *Percept. Psychophys.*, 12:201-204.

Eriksen, C. W. and Hoffman, J. E. (1973). The extent of processing noise elements during selective encoding from visual displays. *Percept. Psychophys.*, 14:155-160.

Eriksen, C. W. and St. James, J. D. (1986). Visual attention within and around the field of focal attraction: a zoom lens model. *Percept. Psychophys.*, 40:225-240.

Frith, C. D., Friston, K., Liddle, P. F. and Frackowiak, R. S. J. (1991). Willed action and the prefrontal cortex in man: A study with PET. *Proc. Roy. Soc. Lond. B.*, 244:241-246.

Harris, L. R. and Jenkin, M. (1998). Vision and action. In L. R. Harris and M. Jenkin (Eds), *Vision and Action*, (pp. 1-12). New York, USA: Cambridge University Press.

Hillstrom, A. P. and Yantis, S. (1994). Visual motion and attentional capture. *Percept. Psychophys.*, 55:399-411.

Jonides, J. and Yantis, S. (1988). Uniqueness of abrupt visual onset in capturing attention. *Percept. Psychophys.*, 43:346-354.

Kapralos, B., Jenkin, M., Milios, E. and Tsotsos, J. (2000). Eyes 'N Ears: A System for Attentive Teleconferencing. Proc. 139th meeting of the Acoustical Society of America, Atlanta, GA.

Mack, A. and Rock, I. (1998). *Inattentional Blindness*. Cambridge, USA: MIT Press.

McCormick, P. A. (1991). Orienting without awareness. *J. Exp. Psychol. Hum. Percept. Perf.*, 23:168-180.

Moore, C. M., Yantis, S. and Vaughn, B. (1998). Object-based visual selection: evidence from perceptual completion. *Psychol. Sci.*, 9:104-110.

Moray, N. (1959). Attention in a dichotic listening: affective cues and the influence of instructions. *Q. J. Exp. Psych.*, 11:56-60.

Motter, B. C. (1998). Neurophysiology of visual attention. In R. Parasuraman (Eds), *The Attentive Brain.* (pp. 51-70). Cambridge, Mass: MIT Press.

Palmer, S. E. (1999). *Vision Science.* Cambridge, MA: MIT Press.

Perrett, D. I., Potter, D. D., Smith, P. A. J., Milner, A. D., Head, A. S., Jeeves, M. A. and Mistlin, A. J. (1985). Visual Cells in the Temporal Cortex Sensitive to Face View and Gaze Direction. *Proc. Roy. Soc. Lond. B.*, 223:293-317.

Posner, M. I. (1980). Orienting of attention. *Q. J. Exp. Psychol.*, 32: 3-25.

Posner, M. I., Nissen, M. J. and Ogden, W. C. (1978). Attended and unattended processing modes: the role of set for spatial locations. In H. L. Pick and B. J. Saltzman (Eds), *Modes of Perceiving and Processing Information.* (pp. 137-158). Hillsdale, NJ.: Erlbaum.

Posner, M. I. and Petersen, S. E. (1990). The attention system of the human brain. *Ann. Rev. Neurosci.*, 13:25-42.

Regan, D. M. (2000). *Human Perception of Objects.* Sunderland, MA: Sinauer.

Rieke, F., Warland, D., de Ruyter van Steveninck, R., and Bialek, W. (1997). *Spikes: Exploring the Neural Code.* Cambridge, USA: MIT Press.

Theewes, J. (1990). Perceptual selectivity is task-dependent: evidence from selective search. *Acta. Psychol.*, 74:81-99.

Treisman, A. (1960). Contextual cues in dichotic listening. *Q. J. Exp. Psych.*, 12:242-248.

Treisman, A. (1998). Feature binding, attention and object perception. *Phil. Trans. R. Soc. Lond. B Biol. Sci.*, 353:1295-1306.

Treisman, A. M. and Gelade, G. (1980). A feature integration theory of attention. *Cog. Psychol.*, 12:97-136.

Tsal, Y. (1983). Movement of attention across the visual field. *J. Exp. Psychol.*, 9:523-530.

Weber, T. A., Kramer, A. F. and Miller, G. A. (1997). Selective processing of superimposed objects: An electrophysiological analysis of object-based attentional selection. *Biol. Psychol.*, 45:159-182.

Wurtz, R. H. and Mohler, C. W. (1976a). Organization of monkey superior colliculus: enhanced visual response of superficial layer cells. *J. Neurophysiol.*, 39:745-765.

Wurtz, R. H. and Mohler, C. W. (1976b). Enhancement of visual responses in monkey striate cortex and frontal eye fields. *J. Neurophysiol.*, 39:766-772.

Yantis, S. (1998). Objects, attention and perceptual experience. In R. Wright (Ed), *Visual Attention*. (pp. 187-214). New York: Oxford University Press.

Yantis, S. and Hillstrom, A. P. (1994). Stimulus driven attentional capture: evidence from equiluminant visual objects. *J. Exp. Psychol.: Hum. Percept. Perform.*, 20:95-107.

Yantis, S. and Jonides, J. (1984). Abrupt visual onsets and selective attention: evidence from visual search. *J. Exp. Psychol.: Hum. Percept. Perform.*, 10:601-621.

Yarbus, A. L. (1967). *Eye Movements and Vision*. New York, USA: Plenum.

Zeki, S. M. (1978). Functional specialisation in the visual cortex of the rhesus monkey. *Nature*, 274:423-428.

2

Shifts of Attention and Saccades Are Very Similar. Are They Causally Linked?

Sally McFadden
Josh Wallman

Attention is first and foremost the servant of cognition, choosing one object among the many available, to paraphrase William James. In addition, attention has a motoric function in selecting the object to be grasped by eye or hand. In this chapter we will first review some of the literature dealing with how visual attention is shifted about and what role it might play in the control of saccadic eye movements, and then briefly describe some experiments of our own on these matters.

2.1   Spatial Attributes of Attention

Directing attention to one location involves paying less attention to other locations. This can be restated as attention facilitating optimization of a limited capacity neural system by limiting the portion of the visual universe with which the nervous system must deal. Thus, the popular spotlight metaphor originally introduced by Posner (1980) involves the concentration of our (limited) neural resources on an object or area. Typically, reaction times (RT's) are found to be faster and visual detection and discrimination are enhanced within an attentional window, with a decrease in performance outside this window (Bashinski and Bacharach, 1980; Eriksen and Yeh, 1985; Jonides, 1983; Posner et al., 1980; Remington, 1980; Steinman et al., 1995).The typical paradigm involves presenting a cue at the same location as a subsequent stimulus (Eriksen and Hoffman, 1972; Posner et al., 1980). Performance is measured at various stimulus onset asynchronies (SOA) between the cue and the subsequent stimulus. This spatial cueing affects performance of a variety of tasks ranging from simple detection to discrimination of shape, size, colour, and motion (Bashinski and Bacharach, 1980; Humphreys, 1981a; Downing, 1988), indicating that processes that mediate spatial attention facilitate many different visual processes. According to a systematic study of the effects of attention in the face of added amounts of noise, this facilitation occurs by reduction in the internal noise of the system or by signal enhancement (Lu and Dosher, 1998).

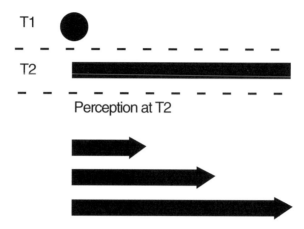

FIGURE 2.1. The line motion illusion. A line presented at time T2, shortly after a transient cue has been presented at time T1, appears to "shoot" or "grow" from the side where the cue has been.

Beyond the effects on psychophysical thresholds, attention appears to reduce the time it takes to process sensory stimuli (Posner, 1980). This effect is embodied in the predominant explanation for the line-motion illusion (Hikosaka et al., 1993). In this illusion, if a line is presented just after a cue located at one end of the line, the line appears to grow from the cued end (Fig. 2.1). One can account for this perceived motion by saying that the attended location has speeded visual processing, causing one end of the line to be perceived before the rest of the line. Although this interpretation has been criticized as ignoring purely visual interactions that might account for the line motion (Downing and Treisman, 1997), the finding that the line-motion illusion is still seen when attention is drawn to one end of the line by auditory or somatosensory stimuli (Shimojo et al., 1997) suggests that at least a substantial part of the effect is attentional.

The behavioural enhancement effects of attention are also reflected at the cellular level because focal attention modulates receptive-field properties in several visual cortical areas, as well as in parietal areas and in the pulvinar (Bushnell et al., 1981; Robinson et al., 1995: Colby et al., 1996; Luck et al., 1997; Kastner et al., 1999; Treue and Maunsell, 1999). Spatially mapped attentional modulation can be detected with fMRI, even within V1 itself (Brefczynski and DeYoe, 1999). This facilitation could reflect either direct enhancement of neural sensitivity at an attended location or could be a consequence of motor preparation for foveating that location.

Early studies claimed that the size of the "spotlight" was typically about 1 deg (Eriksen and Hoffman, 1972; Eriksen and Hoffman, 1973; Humphreys, 1981b) approximately 250 msec after cue onset. However, the spatial distribution of attention has been repeatedly noted to be affected by the acuity demands of the task (Shepherd and Müller, 1989). Furthermore, the extent of the attended area can be modified by varying the probability or expectancy of the target occurring in a particular location (LaBerge, 1983; Müller and Findlay, 1987). These and similar studies typically used variations of the classic cost/benefit approach and were generally directed at measuring the deployment of attentional resources (costs) in situations with more than one stimulus dimension (see review by Kinchla, 1992). The visual search paradigm (Treisman and Gelade, 1980; Treisman, 1977, 1982, 1988) is one such example. These studies do not explicitly address the issue of the spatial extent of the focus of attention, but rather seek to understand the mechanisms underlying shared, divided, or distributed states of attention.

The "zoom-lens" model of attention (Eriksen and Yeh, 1985; Eriksen and St. James, 1986) posits that the size of the attended area can vary in different situations. In this model, attention is characterised as changing gradually from a wide-focus to a narrow "beam" concentrated on the cue. However, this model's prediction that attentional effects would be stronger when the focus is narrow than when broad fails to be supported (Shepherd and Müller, 1989). Other evidence suggests that the spatial scale of attention can change rapidly around the time of saccades (Maljkovic, 1999).

Regardless of the size of the "spotlight" at any particular point in time, there remains the perplexing issue of how attention is allocated in the first place. If attention can be allocated to a particular location, how does this occur? First, a particular location would need to be selected; second, attention would need to be allocated or directed to the selected location (which may involve disengagement and engagement processes); third, some kind of feedback may be generated so that either attention itself or the systems that rely on attention, "know" whether attention has indeed been allocated to the selected location.

2.2   Coordinate Space of Focal Attention

If we speak of focal attention occupying a certain location or area of space, it becomes important to ask to what is the attention attached and in what coordinate space does it lie. There is evidence that when attention is drawn to a brief luminance transient on a moving object, attention continues to move with the object, implying that the attention is (or can be) attached to the object, not the location to which attention had been attracted (Hikosaka et al., 1996). How one views the relation of shifts of attention to eye movements depends on the coordinate space of attention - specifically, whether the map over which attention moves is fixed to the retina or fixed to external space. If focal attention exists in a retinotopic space, it would be moved whenever the eyes moved, requiring it to move in a compensatory direction for it not to lose the attended target. An experiment that tested this distinction involved having a cue flash just before a saccade and assessing the location of attention just after the saccade. The result was that attention seemed to have remained at the cue location, rather than having moved with the eyes (Hikosaka et al., 1996, but see Posner and Cohen 1984). This finding implies a spatial coordinate system, in which the location of attention is uninfluenced by eye movements.

2.3   Overt and Covert Orientation

How do we select a particular location to attend? Normally, in daily life, we look at what we wish to see. Approximately 90% of our viewing time is spent fixating with our fovea items and locations of interest. Fixations are separated by saccadic eye movements, which work to foveate items of interest. Under normal viewing conditions, attention and saccadic eye movements work together to select targets. Attentional allocation that is accompanied by saccadic eye movements is termed *overt orientation*.

Although shifts of visual attention are normally accompanied by eye movements and foveation, one can certainly "look out of the corner of one's eye" (Wundt, 1912). As James (1890) and Helmholtz (1910) noted, we can attend to peripheral locations of interest without moving our eyes. This is termed *covert orientation*. Such covert attentional shifts are very much faster than the associated overt shifts (see later) and may therefore represent the pure attentional changes dissociated from the neural systems underlying eye, head, or other overt body movements. What then are the characteristics of these covert attentional shifts and how do they interact with overt orientation?

2.4   Top-Down versus Bottom-Up Attentional Control

Attention can be summoned in two ways: One can choose to attend to something based on cognitive expectations and elicited through an endogenous (goal-directed) process. This is variously termed endogenous (Posner, 1980) or voluntary (Jonides, 1981; Müller and Rabbitt, 1989), or centrally cued attention. Alternatively, an object can draw attention to itself simply through the sensory properties of the image itself. This is termed exogenous, automatic, reflexive, or peripherally cued (see review by Egeth and Yantis, 1997).

This distinction first arose from a widely used method for inducing a covert attentional shift in which a subject is cued concerning the likely location of the subsequent stimulus. If the time between the cue and the stimulus is short enough, eye movements are precluded and any enhancement of processing the target is attributed to a covert shift of attention. When the cue is presented at a peripheral location, attention shifts to there much more quickly (Eriksen et al., 1990) than when the cue is a number or arrow at the point of fixation indicating the location to which attention should shift (Yantis and Jonides, 1990). The latency of exogenous attentional shifts is less than 100 msec (Remington, 1980; Eriksen, 1990; Yantis and Jonides, 1990), and the associated attentional focus never rests long in one spot, whereas the latency of endogeneous shifts is 300 msec or more (Hikosaka et al., 1996) and can be sustained (Nakayama and Mackeben, 1989). More recently, event-related fMRI analysis has shown distinct brain areas associated with the two sources of orientating (Corbetta, 1998; Coull et al., 2000).

Although there is some general consensus about cost and benefits and ways of eliciting spatially focused attention, there remains lots of uncertainty about how covert shifts in visual attention, whether initiated voluntarily or through the sudden onset of a peripheral transient, occur.

2.5   Shifting Attention

Shifts of attention resemble eye movements. Attention can either move rapidly, but only after a substantial latency, like an eye making saccades, or it can move slowly to follow a moving object, like an eye doing pursuit (Cavanagh, 1992; Hikosaka et al., 1996). Furthermore, there is intriguing evidence that something resembling the oculomotor efference copy signals exists for attentional pursuit. In the oculomotor system, the perceived motion of a target does not cease when it is tracked by the eyes, presumably because an oculomotor efference copy signal contributes to the motion percept (Helmholtz, 1910). In attentional pursuit, when a rotating colored grating is tracked by attention, the speed of the grating is estimated much more accurately than when it is not tracked by attention. Furthermore, when the grating is overlaid by a luminance grating moving in the opposite direction, the motion of the colored grating is only apparent when it is tracked by attention (Cavanagh, 1992). Although other ways of explaining these findings may exist, the experiments suggest that an attentional efference copy signal may exist. Monitoring this signal could supplement the visual information available to the oculomotor system about objects and their motions.

Does this analogy between movements of attention and of the eyes also mean that attention always moves continuously, tracing a trajectory from one point to the next, as the eye must?

Early analog models proposed that attention did indeed move continuously across the visual field (Shulman et al., 1979) at a rate of 8 msec/degree (Tsal, 1983). Tsal's study was based on the slope and changes in the RT-SOA functions for targets located at various eccentricities. However, the distance effects disappear when the known facilitatory warning effects of the cue are controlled for (the so-called "neutral condition"; Remington and Pierce, 1984). The limitation of these studies is that the SOA-RT functions compound the time required to perceive the cue, the latency to initiate a change in attentional focus and the time necessary for the change (Eriksen and Murphy, 1987). A detailed analysis of these studies by Yantis (1988) concluded that they do not provide evidence either for or against an analog model.

Other results argue against the notion that attention moves at a measurable speed. Hughes and Zimba (1985, 1987) found that detection costs, although greater for unexpected signals that occurred in the unattended hemifield, varied independently of the distance separating the expected and unexpected locations (but see Egly and

Homa, 1991). Other studies which controlled for acuity and used discrimination (Sagi and Julesz, 1985) or same–different judgments of rotated letters (Kwak et al., 1991), have also provided evidence for an abrupt relocation of attention that is independent of distance or intervening obstacles (Sperling and Weichselgartner, 1995).

Neural spatial saliency maps (Koch and Ullman, 1985) and the importance of temporal correlation between neurons (Niebur and Koch, 1994) are emphasized in a computational model of the neural basis of selective attention. This is supported by evidence that switching selective attention involves changes in temporal synchrony between neurons (Steinmetz et al., 2000). In this study, extracellular recordings were made in the SII cortex from pairs of active neurons while monkeys switched attention between a visual detection and a tactile task. Although the tactile stimulus was identical in both tasks, when monkeys switched their attention to the tactile task, synchronization increased in 28% of the neuron pairs and decreased in 7%. If attention is represented by the increased synaptic efficacy resulting from an active cooperative mechanism such as synchronicity, then shifting spatial attention would relate to what might trigger synchronous firing in one population of neurons rather than another. Thus, the reallocation of attention would not necessarily be related to retinal or cortical distance per se but rather to the temporal delay that a particular stimulus set—whether derived by exogenous or endogenous means—takes to summon attention, the size of the population of neurons involved, and the ease with which threshold levels of synchronicity in that population can be attained.

2.6   Coupling Between Saccadic Eye Movements and Attentional Shifts

Shifts of attention precede saccadic eye movements. Is there evidence of a causal connection between them? At one extreme, attention and eye movement generation could be construed as completely independent: oculomotor and attentional targeting could be generated in separate spatial maps. Such independence would imply that one could generate a saccade to one target while attending to another location. Alternatively, the programming of shifts of attention and eye movements may be separate processes, but functionally related in that they could be recruited together depending on the task demands. At the other extreme, the same system could control spatial attentional shifts and saccadic shifts.

One model that proposes a connection between saccades and visual attention requires that attention be "disengaged" from the current fixation point for a saccade to occur in response to a new target (Fisher, 1993). This idea explained an observation of Saslow (1967) that saccadic latency can be substantially reduced if the target that the eye is fixating is removed prior to the presentation of a novel target. These "express saccades," defined by their extremely short reaction times, do not occur when attending to either a fixation point or to any other visual stimulus in the peripheral field of view including the saccade target (Fischer and Breitmeyer, 1987). Furthermore, the frequency of express saccades decreases when the saccade target occurs in the close vicinity of a peripheral attention target (Fischer and Weber, 1988). Some authors hold that express saccades occur if visual attention has already been disengaged at the moment when the saccadic target appears (Mayfrank et al., 1986), whereas others argue that the shortened saccadic latencies occur when the foveal stimulus is extinguished whether or not the stimulus had been attended (Kingstone and Klein, 1993; Tam and Stelmach, 1993). It is also possible that express saccades represent eye movements that are planned in anticipation to, rather than as a consequence of, target onset (Husain, 1996).

It has been proposed that the focus of attention moves to the target of a saccadic eye movement during the planning stage of a saccade and prior to its initiation (Hoffman, 1975; Henderson, 1992). Certainly, covert attentional shifts can precede eye movements: Attentional shifts can occur within as little as 60–80 msec (Eriksen et al., 1990), whereas normally the latency of saccades is about 200 msec. A well-supported view is that attention needs to be engaged at a target location before a saccade can be made to that location (Wurtz and Mohler, 1976). Although under certain conditions endogenous attentional shifts can move in a way that does not appear to affect eye-movement preparation (Posner, 1980), in general it is not possible to attend to one location while moving ones eyes to a different one (Deubel and Schneider, 1996). For example, if subjects are required to make a saccade to a specified target while also detecting a visual cue presented just prior to the eye movement, superior cue detection and faster reaction time (Posner et al., 1980; Shepherd et al., 1986) and superior identification performance (Kowler et al., 1995; Chelazzi et al., 1995; Deubel and Schneider, 1996) occur at the saccadic target location regardless of attention instructions (Hoffman and Subramaniam, 1995), even with prior knowledge of the discrimination target position (Deubel and Schneider, 1996). Further evidence that saccades require focal attention, or at least that the two processes coexist, comes from the similar effect of priming on saccade latency and focal attentional deployment in visual-search tasks (McPeek et al., 1999).

Despite this evidence that saccades demand focal attention, some attention can be diverted from the saccade goal, indicating that there is a ceiling on the attentional demands of saccades (Kowler et al., 1995). Kowler suggests that perceptual attention determines the endpoint of the saccade, but proposes that a separate trigger signal initiates the saccade in response to transient changes in the attentional locus.

A more extreme model of the interaction between attention and saccades is represented by the oculomotor readiness theory of Klein (1980). This theory in its later form proposes that attention is tightly linked to exogenously directed eye-movement preparation (Klein, 1992). In a similar, although more extreme, vein, the premotor theory of Rizzolatti (1983) has posited that the system that controls spatial attention is the same one that controls action. Rizzolatti et al. (1987) examined the costs associated with attending to targets at unexpected locations and found hemifield inhibitory effects similar to those of Hughes and Zimba (1985, 1987).

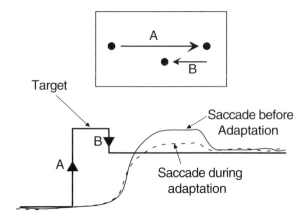

FIGURE 2.2. Saccadic adaptation. A spot of light (black circle) is displaced from the fixation point (by movement A) and is tracked by a saccadic eye movement. However, if the position of the spot is consistently displaced again (by movement B) when the eye is in flight to the original target location, the primary saccade amplitude gradually reduces and becomes recalibrated towards the backstepped location. The boxed area shows the location of the target on the screen, while the associated traces of the target and eye movement are shown below.

The premotor theory attributes typical meridian crossing costs to a mechanism that controls both eye and attention movements. Rizolatti et al. (1987) propose that covert attentional shifts lead to an activation of exogenous oculomotor circuits, even though an eye movement may not actually occur (Sheliga et al., 1994, 1995). In short, they propose that reflexive movements of both attention and saccades are mediated by the same neural circuits.

Attention can also influence the trajectories of saccades. Saccades produced by electrical stimulation of the superior colliculus in monkeys are influenced by attention (Kustov and Robinson, 1996), and in humans saccade trajectories are affected by the location of attention in that the allocation of attention to the upper (or lower) hemifield causes a downward (or upward) saccade deviation respectively (Sheliga et al., 1995).

Thus, much evidence points to significant coupling between attentional shifts and the saccadic eye movements, although there remains disagreement as to the nature of this coupling. Overall, the evidence supports an attentional mechanism that may guide the saccadic system in the selection of a particular spatial location as well as provide the information necessary for motor action. We will now consider how saccades are controlled and how attention might fit into this control process.

2.7   Adaptive Control of Saccadic Eye Movements

Saccadic eye movements cannot be visually guided because their speed is too great and their duration too short for effective on-line visual processing. Any system, whether biological or man-made, that moves in an open loop manner (i.e., without continuous guidance) must have some way of recalibrating the magnitude of its output relative to its input. Otherwise, it will gradually drift towards making movements that are either too large (hypermetric) or too small (hypometric).

The accuracy of saccades is controlled by adaptive mechanisms (see review by Optican, 1985). If the oculomotor system is deceived into thinking that it is making hypermetric saccades, adaptive changes in the saccade amplitude occur gradually over many saccades. This was demonstrated originally by McLaughlin (1967), who stepped a target (a small spot of light) and while the saccade to the target was under way, surreptitiously displaced the target backward (Fig. 2.2). When the eye landed after the saccade, it appeared that the saccade had been too large. Over trials, the size of the primary saccade gradually declined so that the saccades landed progressively closer to the displaced position of the target rather than to the original position of the target before the saccade. These adaptive changes in saccade amplitude were quite specific to the experimental situation. Thus, saccades can be adapted independently for opposite directions of movement (that is, rightward saccades can be increased while leftward ones decrease; Semmlow et al., 1989) and for different amplitudes (that is, 5° saccades can be increased while 20° saccades are decreased; Miller et al., 1981). Different types of saccades can be adapted independently, for example, visually triggered saccades can be increased without affecting memory-guided saccades or free-scanning saccades (Erkelens and Hulleman, 1993; Deubel, 1995).

The motor learning seen in saccadic adaptation probably functions to keep the saccades reasonably accurate in the face of changes in eye-size during development, changes in muscle strength or changes in neural efficacy. When one eye-muscle is deliberately weakened by surgery, the recovery of normal saccade function is similar to the changes provoked by the saccade-gain adaptation experiments described earlier (Scudder et al., 1998).

2.8   Nature of the Error Signal

What is it about finding the target in an unexpected location after a saccade that produces saccadic gain adaptation? One possibility is that the brain cross-correlates the visual information around the target before and after the saccade. If the scene that was around the target is not consistently centered on the fovea after the saccade, a need for gain correction would be indicated. Evidence that such correlations can be used by the saccadic system without conscious knowledge comes from an experiment in which the saccadic target was defined by the transient caused by the reversal of a part of a complex one-dimensional random grating. During

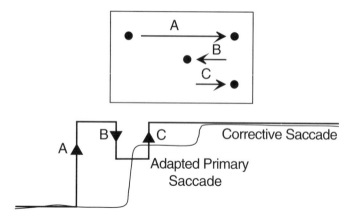

FIGURE 2.3. Double step paradigm of Deubel. As is in the saccadic adaptation paradigm (Fig. 2.2), a small target is backstepped by distance B while the eye is in flight in response to the initial step A. If shortly thereafter, the target is stepped forward again (by movement C), the eye does not make a corrective saccade to the backstepped location. Under these conditions, the saccade amplitude still adapts towards the backstepped location.

the ensuing saccade, if a part of the grating including the "target" was moved, a corrective saccade followed, even though the target movement was not noticed by the subject (Deubel et al., 1984).

Alternatively, the oculomotor system might not need to know which target the saccade was directed toward in order to judge the accuracy of saccades. Instead, it could simply keep track of whether it needs to make a corrective saccade and, if so, in what direction. While attractive in its simplicity, this hypothesis cannot account for the fact that one can arrange matters so that the corrective saccades are forward (as though the saccade were too small) but the saccadic gain decreases (as though the saccade were too large). One does this by having the target step backward during the saccade, but, before the subject has a chance to make a corrective saccade, the target returns to its original location (double-step paradigm, Fig. 2.3). In this situation, the saccade gain decreases, with the result that most of the corrective saccades are in the same direction as the primary saccade (Wallman and Fuchs, 1998).

2.9   Are Shifts of Attention also Adaptable?

Shifts of attention are like saccades in that the locus of attention moves very swiftly (perhaps, as indicated earlier, instantaneously) after a considerable delay. Whether or not one holds that the mechanism of shifts of attention is intrinsically related to the mechanism of saccades, saccades are generally made to the locus of attention. Therefore, by analogy with changing the magnitude of saccades gradually by tricking the oculomotor system into correcting an apparent mismatch between saccade size and target distance, we tested the possibility that the magnitude of

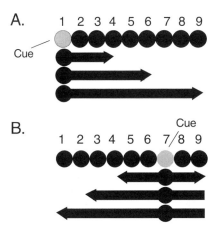

FIGURE 2.4. Locating attention. In a variation of the line-motion illusion, we divided the line into nine circles, and trained subjects to accurately locate the origin of the observed motion. Two examples are shown in which a cue (identical to one of the circles) was briefly flashed prior to the presentation of the line made of the row of circles. (a) The cue located at position 1 gives the perception of a strong motion vector directed away from the position 1. (b) When the cue is located at intermediate positions, such as position 7, rightward motion is also seen, but leftward motion predominates. When the cue is located at position 5 (not shown), the motion appears to shoot equally in both directions.

shifts of attention can be similarly manipulated.

To do this, we, together with Afsheen Khan, explored whether shifts of attention could have their magnitude changed by essentially the same technique used to modify the magnitude of saccades. With the subject fixating a stationary point, we stepped a cue and then, at the time that we estimated attention was shifting to the cue, we moved the cue back a little so that the attentional shift would appear to have been systematically too large and we looked for a gradual change in the size of the attentional shift.

What makes this experiment difficult to carry out is, first, that it is not easy to know at which moment attention is shifting, and second, that it is not easy to know the magnitude of the attentional shift. We addressed the first problem by using the line-motion illusion (Hikosaka et al., 1993), described earlier, to determine the average latency for each subject to shift attention, and used the latency obtained in this way to determine when to step the cue back. To find the attention-shift latency for each subject, we had a cue appear 10 deg from the fixation point and, at varying intervals afterwards, we presented a line spanning the distance between fixation point and cue. Subjects reported that the line appeared to grow from the fixation point at short intervals and from the cue at long intervals. We used the 50% interval as the attention-shift latency.

The next problem was how to identify where the attention moved to when the cue appeared. To do this we used a variant of the line-motion illusion in which we replaced the line by a row of nine circles. After training subjects to type a number

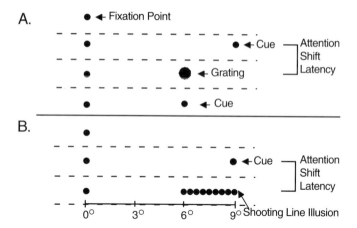

FIGURE 2.5. Sequence of events to adapt attention. The majority of trials (87%) were used to adapt attention (a) and were interleaved with trials that were used to measure where attention landed (b). Subjects were required to continuously foveate the fixation spot located in the centre of the screen and shift attention to a peripherally located target cue. At each subject's attentional shift latency, the cue was stepped back and either (a) replaced by a small grating or (b) replaced by the row of circles. The subject's task was either to discriminate the orientation of the grating stimulus (a) or report the origin of the shooting line (b).

on a keyboard that represented which of the circles had been briefly illuminated, we arranged to have the cue appear at various locations along the row of circles and trained subjects to identify accurately the circle from which the line appeared to emerge (Fig. 2.4). The origin of the shooting line indicated the location of the subject's attention at the time when the line (row of circles) was presented.

During the adaptation experiment, the subject's task was to fixate on a centrally located cue and track a small peripheral cue with covert attention. Two different types of task were interleaved (Fig. 2.5). On most trials, the cue appeared to the left or right of the fixation point and then, after the attention-shift latency, the cue was stepped back by 30% and was then briefly replaced by a small grating, the orientation of which the subject identified (Fig. 2.5a). Although the cue onset attracted attention exogenously to the backstepped location, the grating task maintained attention endogenously at this position. The second task occurred on 13% of the trials (Fig. 2.5b), and here the cue did not step back but instead was replaced by the line of circles. In this task, the subject was required to identify the circle from which the line emerged, providing a measure of where attention landed.

We found that over the course of 600 trials, the position from which the line appeared to emerge moved progressively in the direction of the backstepped location, implying that the magnitude of the shift of attention to the cue had decreased as a result of the adaptation. Subjects generally showed a decrease in the magnitude of the shift of attention, the average being a decrease of about 25% of the backstep. It is not surprising that the magnitude of the attention shift was smaller than with saccadic adaptation (generally about 65%), given that in most cases the backstep must have occurred either slightly before or slightly after the actual shift of attention, unlike the situation in saccadic-adaptation experiments in which one can be certain that the backstep occurs exactly at the time of the saccade.

One of the striking results of this experiment was that the adaptation always occurred gradually over many trials. One might think that one could learn where the cue was going to backstep to and direct attention to that locus. No subject did that. We suspect the reason is that the saliency of the cue always attracted attention by exogenous mechanisms and that this is what we measured. The top-down endogenous mechanisms act more slowly and could only direct attention long after we had made our measurement of the locus of attention. We consider this as evidence that we are studying an adaptational process rather than a cognitive strategy.

We also found that this adaptation of the shifts of attention is specific to the visual hemifield and that the magnitude of the shifts of attention can be increased as well as decreased. To show this, we adapted subjects under conditions in which, when the cue appeared on the right, it stepped back after the attention-shift latency, but when the cue appeared on the left it did not step back. In all three subjects, the magnitude of the shifts of attention decreased for rightward shifts but not for leftward shifts. In other experiments, if we had the cues step forward, rather than back at the subject's attention-shift latency, the subjects showed an increase in the magnitude of the attentional shifts. These characteristics of the adaptation are also seen in saccadic adaptation.

The fact that one can adapt shifts of attention is somewhat surprising. Either it implies that shifts of attention are "motoric" in some sense and require calibration, like any open-loop behavior, or that experience is required to maintain the alignment of the map of attention with the visual map.

Furthermore, if saccades elicited by peripheral transients are actually directed to the locus to which attention has shifted, the fact that this locus is adaptable suggests that what is called saccadic adaptation might, in some cases, actually be adaptation of attention. To investigate whether the adaptation of attention and of saccades might be linked, we measured saccadic amplitude before and after adapting attention in both directions. In five of seven subjects, the saccadic amplitude was significantly lowered by the attentional adaptation, even though no saccades at all were made during the period of adaptation. On the other hand, in three subjects in which we adapted shifts of attention in only one direction, no saccadic adaptation was observed, so this matter remains uncertain at present.

2.10   Might Attention Provide an Error Signal to Saccade-Gain Adaptation?

The close linking of attention and saccades suggested to us the possibility that attention might have another function in saccadic programming, as an error signal in saccadic adaptation. It is attractive to consider saccadic-gain adaptation to operate close to the level of the saccade generator rather than at a higher, more cognitive level. If this is the case, it is puzzling how the gain controller knows which of the multitude of targets a given saccade was made to, knowledge that would seem necessary to decide whether the saccade was accurate or was too large or too small. Two possible solutions have already been mentioned. The saccadic-gain controller could simply observe whether saccades were followed by corrective saccades in the same or opposite direction (motoric error signal). Alternatively, the visual system could cross-correlate the spatial pattern of visual stimuli near the target before and after the saccade. Another possible solution would be to regard any stimulus near the fovea after a saccade as the target, so that if there were more salient visual stimuli on the near side of the fovea after a saccade, the saccade would be considered too large. We propose a fourth solution, that after a saccade, the locus of attention might provide an error signal. Because saccades are generally to targets that attract attention, if after a saccade attention is drawn consistently to the near side of the fovea, the saccadic-gain controller could infer that the target of the saccade is now on that side, an indication that the saccade was too large.

To explore the plausibility of either the visual or attentional hypothesis, we, together with Raymond Yun, arranged for saccades to be made to targets that disappeared for 300 msec, beginning while the saccade was in flight. In their place, we produced an entirely different stimulus (either a whirling vane pattern, a contracting circular grating, a moving field of dots, or a dynamic noise pattern) on the near side of the target location. These patterns, designed to attract attention, resulted in significantly decreased saccadic gain, even though the target of the saccades was never elsewhere from where it had been before the saccade (Figure 2.6). This finding can be interpreted as evidence for either a visual or attentional error signal. Alternatively, one might imagine that the saccadic gain control mechanism simply regarded the attractor, which replaced the target, as the target itself and thus reacted as it would to a target consistently displaced during saccades.

To distinguish between these two interpretations, we did an experiment in which the target did nothing unusual: It neither backstepped, nor disappeared, but simply stepped back and forth on the screen. However upon each saccade that the subject made, we added one of the jazzy attractors just described consistently on one side of the target. In this situation as well, the saccade gain was decreased if the attractors were on the near side of the target and increased if the attractors were on the far side of the target. Although the gain changes were very small, most were statistically significant. (Gain changes were evaluated by comparing a set of saccades before and after the adaptation trials.) We interpret these experiments as evidence that either attention, or the visual stimuli that attract attention, can act as an error signal to the saccadic-gain controller.

The results of the experiments just discussed suggest two roles for attention in saccadic-gain control. The magnitude of shifts of attention itself can be modified and thereby may be able to modify the location to which saccades are made because of the role of attention in saccadic targeting. Entirely independent of this role, attention may provide an error signal to inform the saccadic controller of whether the target to which the saccade was directed is on the near or far side of the fovea

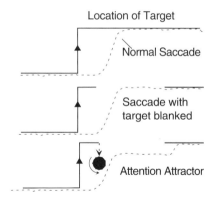

FIGURE 2.6. Effect of attention attractors on saccade amplitude. Saccades (dotted lines) are made to the location of the target (solid lines) whether or not the target is blanked at the time that the saccade is made (upper two panels). If attention is drawn to a nearby location (bottom panel), saccade gain adapts slightly toward the location of the attention attractor, although the eye does not make saccades toward the attractor.

after the saccade.

We speculate that the reason that one can independently adapt different types of saccades may be because one might be adapting shifts of attention in some cases and motoric gains in others. It hardly seems surprising that attention and saccades might be, on the one hand, linked, because of the role of attention in saccadic targeting, and, on the other hand, independent, because one would need to change the motoric gain of saccades without changing the mapping of attention in the world. After all, saccadic targeting is not the only reason for moving attention around.

References

Bashinski, H. S. and Bacharach, V. R. (1980). Enhancement of perceptual sensitivity as the result of selectively attending to spatial locations. *Percept. Psychophys.*, 28:241-248.

Brefczynski, J. A. and DeYoe, E. A. (1999). A physiological correlate of the 'spotlight' of visual attention. *Nat. Neurosci.*, 2:370-374.

Bushnell, M. C., Goldberg, M. E., and Robinson, D. L. (1981). Behavioral enhancement of visual responses in monkey cerebral cortex. I. Modulation in posterior parietal cortex related to selective visual attention. *J. Neurophysiol.*, 46:755-772.

Cavanagh, P. (1992). Attention-based motion perception. *Science*, 257:1563-1565.

Chelazzi, L., Biscaldi, M., Corbetta, M., Peru, A., Tassinari, G. and Berlucchi, G.

(1995). Oculomotor activity and visual spatial attention. *Behav. Brain Res.*, 71:81-88.

Colby, C. L., Duhamel, J. R. and Goldberg, M. E. (1996). Visual, presaccadic, and cognitive activation of single neurons in monkey lateral intraparietal area. *J. Neurophysiol.*, 76:2841-2852.

Corbetta, M. (1998). Frontoparietal cortical networks for directing attention and the eye to visual locations: identical, independent, or overlapping neural systems? *Proc. Nat. Acad. Sci. USA*, 95:831-838.

Coull, J. T., Frith, C. D., Buchel, C. and Nobre, A. C. (2000). Orienting attention in time: behavioural and neuroanatomical distinction between exogenous and endogenous shifts. *Neuropsychologia*, 38:808-819.

Deubel, H. (1995). Separate adaptive mechanisms for the control of reactive and volitional saccadic eye movements. *Vis. Res.*, 35:3529-3540.

Deubel, H. and Schneider, W. X. (1996). Saccade target selection and object recognition: evidence for a common attentional mechanism. *Vis. Res.*, 36:1827-1837.

Deubel, H., Wolf, W. and Hauske, G. (1984). The evaluation of the oculomotor error signal. In A. G. Gale and F. Johnson (Eds), *Theoretical and Applied Aspects of Eye Movement Research*. Amsterdam: Elsevier.

Downing, C. J. (1988). Expectancy and visual-spatial attention: effects on perceptual quality. *J. Exp. Psychol. Hum. Percept. Perf.*, 14:188-202.

Downing, P. E. and Treisman, A. M. (1997). The line-motion illusion: attention or impletion? *J. Exp. Psychol. Hum. Percept. Perform.*, 23:768-779.

Egeth, H. E. and Yantis, S. (1997). Visual attention: control, representation, and time course. *Ann. Rev. Psychol.*, 48:269-297.

Egly, R. and Homa, D. (1991). Reallocation of visual attention. *J. Exp. Psych. Human Percept. Perf.*, 17:142-159.

Eriksen, C. W. (1990). Attentional search of the visual field. In D. Brogan (Ed.) *Visual Search*, 221-240. London: Taylor Francis.

Eriksen, C. W. and Hoffman, J. E. (1972). Temporal and spatial characteristics of selective encoding from visual displays. *Percept. Psychophys.*, 12:201-204.

Eriksen, C. W. and Hoffman, J. E. (1973). The extent of processing of noise elements during selective encoding from visual displays. *Percept. Psychophys.*, 14:155-160.

Eriksen, C. W. and Murphy, T. D. (1987). Movement of attentional focus across the visual field: a critical look at the evidence. *Percept. Psychophys.*, 42:299-305.

Eriksen, C. W. and St. James, J. D. (1986). Visual attention within and around the field of focal attention: a zoom lens model. *Percept. Psychophys.*, 40:225-240.

Eriksen, C. W. and Yeh, Y. Y. (1985). Allocation of attention in the visual field. *J. Exp. Psychol. Hum. Percept. Perf.*, 11:583-597.

Erkelens, C. J. and Hulleman, J. (1993). Selective adaptation of internally triggered saccades made to visual targets. *Exp. Brain Res.*, 93:157-164.

Fischer, B. and Breitmeyer, B. (1987). Mechanisms of visual attention revealed by saccadic eye movements. *Neuropsychologia*, 25:73-83.

Fischer, B. and Weber, H. (1988). Significance of attentive fixation for the selection of saccade targets in different parts of the visual field of the rhesus monkey. *Exp. Brain Res.*, 73: 577-582.

Fisher, B. W. (1993). Express saccades and visual attention. *Behav. Brain Sci.*, 16:553-610.

von Helmholtz, H. (1910). *Treatise on Physiological Optics* (trans. 1925 from the third German edition, J. P. C. Southall, Ed). New York: Dover (1962).

Henderson, J. M. (1992). Visual attention and eye movement control during reading and scene perception. In K. Rayner (Ed.) *Eye Movements and Visual Cognition*, 260-283. New York: Springer-Verlag.

Hikosaka, O., Miyauchi, S. and Shimojo, S. (1993). Focal visual attention produces illusory temporal order and motion sensation. *Vis. Res.*, 33:1219-1240.

Hikosaka, O., Miyauchi, S. and Shimojo, S. (1996). Orienting of spatial attention‹its relexive, compensatory and voluntary mechanisms. *Cog. Brain Res.*, 5:1-9.

Hoffman, J. E. (1975). Hierarchical stages in the processing of visual information. *Percept. Psychophys.*, 18:348-354.

Hoffman, J. E. and Subramaniam, B. (1995). The role of visual attention in saccadic eye movements. *Percept. Psychophys.*, 57:787-795.

Hughes, H. C. and Zimba, L. D. (1985). Spatial maps of directed visual attention. *J. Exp. Psych. Human Percept. Perf.*, 11:409-430.

Hughes, H. C. and Zimba, L. D. (1987). Natural boundaries for the spatial spread of directed visual attention. *Neuropsychologia*, 25:5-18.

Humphreys, G. W. (1981a). Flexibility of attention between stimulus dimensions. *Percept. Psychophys.*, 30:291-302.

Humphreys, G. W. (1981b). On varying the span of visual attention: Evidence for two modes of spatial attention. *Q. J. Exp. Psych. A, Human Exp. Psych.*, 33A:17-30.

Husain, M. A. K., (1996). The role of attention in human oculomotor control. In W. H. Zangemeister, H. S. Stiehl and C. Freksa (Eds.), *Visual Attention and Cognition*, 116:165-175. Amsterdam: Elsevier.

James, W. (1890). *The Principles of Psychology*, New York: Holt.

Jonides, J. (1981). Voluntary versus automatic control over the mind's eye movements. In A. D. Long (Ed.), *Attention and Performance IX*, 187-203, Hillsadale, NJ: Lawrence Erlbaum Associates.

Jonides, J. (1983). Further toward a model of the mind's eye's movement. *Bull. Psychon. Soc.*, 21:247-250.

Kastner, S., Pinsk, M. A., De Weerd, P., Desimone, R. d Ungerleider, L. G. (1999). Increased activity in human visual cortex during directed attention in the absence of visual stimulation. *Neuron.*, 22:751-761.

Kinchla, R. A. (1992). Attention. *Ann. Rev. Psychol.*, 43:711-742.

Kingstone, A. and Klein, R. M. (1993). Visual offsets facilitate saccadic latency: does predisengagement of visuospatial attention mediate this gap effect? *J. Exp. Psychol. Hum. Percept. Perf.*, 19:1251-1265.

Klein, R. M. (1980). Does oculomotor readiness mediate cognitive control of visual attention? in (R. Nickerson, ed.), *Attention and Performance, Vol. VIII*. Hillsdale, NJ: Erlbaum.

Klein, R., Kingston, A. and Pontefract, A. (1992). Orientating of visual attention. In K. Rayner (Ed.) *Eye Movements and Visual Cognition*, pp. 46-65, New York, NY: Springer-Verlag.

Koch, C. and Ullman, S. (1985). Shifts in selective visual attention: towards the underlying neural circuitry. *Hum. Neurobiol.*, 4:219-227.

Kowler, E., Anderson, E., Dosher, B. and Blaser, E. (1995). The role of attention in the programming of saccades. *Vis. Res.*, 35:1897-1916.

Kustov, A. A. and Robinson, D. L. (1996). Shared neural control of attentional shifts and eye movements. *Nature*, 384:74-77.

Kwak, H. W., Dagenbach, D. and Egeth, H. (1991). Further evidence for a time-independent shift of the focus of attention. *Percept. Psychophys.*, 49:473-480.

LaBerge, D. (1983). Spatial extent of attention to letters and words. *J. Exp. Psychol. Human Percept. Perf.*, 9:371-379.

Lu, Z. L. and Dosher, B. A. (1998). External noise distinguishes attention mechanisms. *Vis. Res.*, 38:1183-1198.

Luck, S. J., Chelazzi, L., Hillyard, S. A. and Desimone, R. (1997). Neural mechanisms of spatial selective attention in areas V1, V2, and V4 of macaque visual cortex. *J. Neurophysiol.*, 77:24-42.

Maljkovic, V. (1999). Spatial scale of attentional deployment changes during saccade preparation. *Invest. Ophthal. Vis. Sci.*, 40:S415.

Mayfrank, L., Mobashery, M., Kimmig, H. and Fischer, B. (1986). The role of fixation and visual attention in the occurrence of express saccades in man. *Eur. Arch. Psychiatry Neurol. Sci.*, 235:269-275.

McLaughlin, S. (1967). Parametric adjustment in saccadic eye movements. *Percept. Psychophys.*, 2:359-362.

McPeek, R. M., Maljkovic, V. and Nakayama, K. (1999). Saccades require focal attention and are facilitated by a short-term memory system. *Vis. Res.*, 39: 1555-1566.

Miller, J., Anstis, T. and Templeton, W. (1981). Saccadic plasticity: Parametric adaptive control by retinal feedback. *J. Exp. Psychol. Hum. Percept. Perf.*, 7:356-366.

Müller, H. J., and Findlay, J. M. (1987). Sensitivity and criterion effects in the spatial cuing of visual attention. *Percept. Psychophys.*, 42: 383-399.

Müller, H. J. and Rabbitt, P. M. (1989). Reflexive and voluntary orienting of visual attention: time course of activation and resistance to interruption. *J. Exp. Psychol. Hum. Percept. Perf.*, 15:315-330.

Nakayama, K. and Mackeben, M. (1989). Sustained and transient components of focal visual attention. *Vis. Res.*, 29:1631-1647.

Niebur, E. and Koch, C. (1994). A model for the neuronal implementation of selective visual attention based on temporal correlation among neurons. *J. Comput. Neurosci.*, 1:141-158.

Optican, L. M. (1985). Adaptive properties of the saccadic system. In A. M. J. Berthoz (Ed.) *Adaptive mechanisms in gaze control: Facts and theories, Vol. 1*, pp. 71-79. Amsterdam: Elsevier.

Posner, M. I. (1980). Orienting of attention. *Q. J. Exp. Psychol.*, 32:3-25.

Posner, M. I. Snyder, C. R. and Davidson, B. J. (1980). Attention and the detection of signals. *J. Exp. Psychol.*, 109:160-174.

Posner, M. L. and Cohen, Y. (1984). Components of visual orienting. In H. Bouma and D. Bouwhuis (Eds.) *Attention and Performance X*, pp. 531-556. London: Erlbaum.

Remington, R. (1980). Attention and saccadic eye movements. *J. Exp. Psychol. Hum. Percept. Perf.*, 6:726-744.

Remington, R. and Pierce, L. (1984). Moving attention: Evidence for time-invariant shifts of visual selective attention. *Percept. Psychophys.*, 35:393-399.

Rizzolatti, G. (1983). Mechanisms of selective attention in mammals. In J. P. E-wart, R. R. Capranica and D. J. Ingle (Eds.) *Advances in Vertebrate Neuroethology*, pp. 261-297. New York: Plenum.

Rizzolatti, G., Riggio, L., Dascola, I. and Umilta, C. (1987). Reorienting attention across the horizontal and vertical meridians: Evidence. *Neuropsychologia*, 25:31-40.

Robinson, D. L., Bowman, E. M. and Kertzman, C. (1995). Covert orienting of attention in macaques. II. Contributions of parietal cortex. *J. Neurophysiol.*, 74:698-712.

Sagi, D. and Julesz, B. (1985). Fast noninertial shifts of attention. *Spat. Vis.*, 1: 141-149.

Saslow, M. G. (1967). Effects of components of displacement-step stimuli upon latency for saccadic eye movement. *J. Opt. Soc. Am.*, 57:1024-1029.

Scudder, C. A., Batourina, E. Y. and Tunder, G. S. (1998). Comparison of two methods of producing adaptation of saccade size and implications for the site of plasticity. *J. Neurophysiol.*, 79:704-715.

Semmlow, J., Gauthier, G. and Vercher, J.-L. (1989). Mechanisms of short-term saccadic adaptation. *J. Exp. Psychol. Hum. Percept. Perf.*, 15:249-258.

Sheliga, B. M., Riggio, L., Craighero, L. and Rizzolatti, G. (1995). Spatial attention-determined modifications in saccade trajectories. *Neuroreport*, 6:585-588.

Sheliga, B. M., Riggio, L. and Rizzolatti, G. (1994). Orienting of attention and eye movements. *Exp. Brain Res.*, 98:507-522.

Shepherd, M., Findlay, J. M. and Hockey, G. R. J. (1986). The relationship between eye movements and spatial attention. *Q. J. Exp. Psychol. A*, 38:475-491.

Shepherd, M. and Müller, H. J. (1989). Movement versus focusing of visual attention. *Percept. and Psychophys.*, 46:146-154.

Shimojo, S., Miyauchi, S. and Hikosaka, O. (1997). Visual motion sensation yielded by non- visually driven attention. *Vis. Res.*, 37:1575-1580.

Shulman, G. L., Remington, R. W. and McLean, J. P. (1979). Moving attention through visual space. *J. Exp. Psychol. Hum. Percept. Perf.*, 5:522-526.

Sperling, G. and Weichselgartner, E. (1995). Episodic theory of the dynamics of spatial attention. *Psychol. Rev.*, 102:503-532.

Steinman, B. A., Steinman, S. B. and Lehmkuhle, S. (1995). Visual attention mechanisms show a center-surround organization. *Vis. Res.*, 35:1859-1869.

Steinmetz, P. N., Roy, A., Fitzgerald, P. J., Hsiao, S. S., Johnson, K. O. and Niebur, E. (2000). Attention modulates synchronized neuronal firing in primate somatosensory cortex. *Nature*, 404:187-190.

Tam, W. J. and Stelmach, L. B. (1993). Viewing behavior: ocular and attentional disengagement. *Percept. Psychophys.*, 54:211-22.

Treisman, A. M. (1977). Focused attention in the perception and retrieval of multidimensional stimuli. *Percept. Psychophys.*, 22:1-11.

Treisman, A. M. (1982). Perceptual grouping and attention in visual search for features and for objects. *J. Exp. Psychol. Hum. Percept. Perf.*, 8:194-214.

Treisman, A. M. (1988). Features and objects: the fourteenth Bartlett memorial lecture. *Q. J. Exp. Psychol. A*, 40:201-237.

Treisman, A. M., and Gelade, G. (1980). A feature-integration theory of attention. *Cognit. Psychol.*, 12:97-136.

Treue, S. and Maunsell, J. H. (1999). Effects of attention on the processing of motion in macaque middle temporal and medial superior temporal visual cortical areas. *J. Neurosci.*, 19:7591-7602.

Tsal, Y. (1983). Movements of attention across the visual field. *J. Exp. Psychol. Hum. Percept. Perf.*, 9:523-530.

Wallman, J. and Fuchs, A. F. (1998). Saccadic gain modification: visual error drives motor adaptation. *J. Neurophysiol.*, 80:2405-2416.

Wundt, W. (1912). *Introduction to Psychology*. London: Allen.

Wurtz, R. H. and Mohler, C. W. (1976). Organization of monkey superior colliculus: enhanced visual response of superficial layer cells. *J. Neurophysiol.*, 39:745-765.

Yantis, S. (1988). On analog movements of visual attention. *Percept. Psychophys.* 43:203-206.

Yantis, S. and Jonides, J. (1990). Abrupt visual onsets and selective attention: voluntary versus automatic allocation. *J. Exp. Psychol. Hum. Percept. Perf.* 16:121-134.

3

Contrast Gain, Area Summation and Temporal Tuning in Primate Visual Cortex

Michael J. Hawken
Robert M. Shapley
Michael P. Sceniak
Dario L. Ringach
Elizabeth N. Johnson

3.1 Introduction

Neurones at all levels of the visual pathway have receptive fields that can be described by measuring the neurone's response to stimuli that vary in space and time. The result of this mapping is referred to as the spatiotemporal receptive field. In the retina and the lateral geniculate nucleus (LGN), the spatial receptive field has a center surround organization that is approximately circularly symmetric. For some retinal and LGN cells, the temporal component of the response is sustained: they are low-pass temporal-frequency tuning. Other retinal and LGN cells show transient temporal characteristics; they are bandpass in their temporal-frequency tuning. In the primary visual cortex, there is refinement of some characteristics of the spatio-temporal receptive field and the emergence of new properties (Hubel and Wiesel, 1962, 1968). The sharper tuning for spatial frequency is an example of refinement, whereas the appearance of orientation selectivity is an example of an emergent property. In addition, the temporal characteristics of many neurones are modified compared to their LGN inputs.

Another feature of visual cortical function is the maintenance of tuning over a wide range of input signal strengths. Although visual neurones have a limited firing range, they are able to alter their gain dynamically, thus maintaining sensitivity over a wide range of light levels and contrasts. This second process that dynamically alters gain depending on the ambient stimulus conditions is often called gain-control or normalization. One of the fundamental benefits attributed to gain control mechanisms is that they allow sensory systems to maintain sensitivity while stimulus intensity varies widely and to maintain constant tuning characteristics of stimulus attributes such as size and orientation.

In this chapter, we examine the degree to which the spatiotemporal organization of the receptive field remains constant under different degrees of contrast adaptation. We discuss some of the recent work that suggests that there are quite pronounced changes in the spatiotemporal properties of the cortical receptive field with changes in stimulus contrast. Some of these changes that are observed with changing contrast seem to be altering the fundamental spatiotemporal structure of the receptive fields, which leads us to question the idea of invariance in the spatiotemporal tuning of the receptive field during normalization.

3.2   Gain Control

The visual system operates under severe constraints of dynamic range. One of the mechanisms that allows the neurones to obtain a larger dynamic range is gain control. There are a number of forms of gain control that have been established psychophysically and, for some of these forms, the neural basis has also been established. Probably the best-known gain control is associated with adaptation to different intensities of light. To allow primates to see over a range of illumination that covers many log units (more than 8),two principal mechanisms operate. One is based on the sensitivity of the different classes of photoreceptors (rods and cones), which operate at different levels of light intensity. The rods operate at low light levels (night vision), whereas the cones operate in daylight. The second is due to changes in gain with changes in local illumination. In photopic illumination, where only the cones are operating, primates are able to see spatial form in conditions where the illumination in a scene can vary by many orders of magnitude. This mechanism that maintains sensitivity despite large variations in illumination is called light adaptation. Thus, light adaptation is a dynamic process that lets the visual system operate over ranges of light levels that are greater than could be obtained with the limited dynamic range of neurones.

Psychophysically, observers are most sensitive to local changes in illumination that are close to the mean level of local illumination (Craik, 1938). On a bright sunlit day, we are sensitive to changes in illumination (contrast) of objects that are both directly illuminated and in shadow. Sometimes, illuminated regions are more than two log units darker or brighter than the overall mean level of illumination of the whole scene, yet we are sensitive to objects over the whole scene. This suggests that light adaptation is a local process. Our retina automatically adapts to the ambient illumination level locally, both in the bright and shadowed areas of a scene. The sensitivity to local contrast remains relatively constant despite uneven illumination in the scene. This is achieved in the neural circuits of the retina. The retinal cells alter their gain or level of responsiveness so that they give their largest responses to changes in illumination (contrast) around the local mean illumination level (Shapley and Enroth-Cugell, 1983). The main stimulus to the visual system is local contrast.

FIGURE 3.1. A demonstration of induced contrast changes of a central grating patch by the arrangement of the spatial characteristics of a surrounding patch. The physical contrast of the central patch is equal in all nine panels. The apparent contrast of the central patch is reduced when the surrounding region has the same spatial frequency (top middle), orientation (middle left),or spatial phase (bottom left). The figure illustrates that some of the effects of the surround on the apparent contrast of the center are spatially specific and are likely to be due, in part at least, to mechanisms that are cortical in origin. Reprinted with permission from D'Zmura and Singer (1999).

3.3   Contrast-Gain Control

Most of the signals that arise in the visual pathways do not arise in response to overall changes in the ambient illumination but as responses to changes in local illumination, or as it is more commonly called, spatial contrast. In addition to having mechanisms to alter the sensitivity to local illumination,the visual system also has mechanisms to alter its gain to contrast, mechanisms that are dependent on the level of ambient local contrast (Shapley and Victor, 1978; Shapley and Victor, 1981; Ohzawa et al., 1985; Albrecht and Geisler, 1991; Heeger, 1992; Benardete et al, 1992; Benardete and Kaplan, 1999). This is called contrast-gain control, a phenomenon that has both retinal and cortical components (D'Zmura and Singer, 1999). Contrast gain control is most easily thought of in terms of maintaining an optimal sensitivity to changes in local contrast levels that are close to the mean level of local contrast. Psychophysically, this is illustrated by the

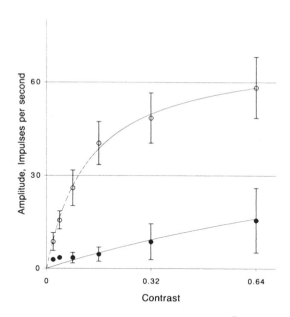

FIGURE 3.2. Average contrast response functions for 36 S potentials recorded from one rhesus monkey: 28 recorded in the parvocellular layers (closed circles) and 8 recorded in the magnocellular layers (open circles). The stimulus was a sinusoidal grating of optimal spatial frequency for each cell, drifting at 4 Hz. The error bars are $\pm$ 1 SD. The smooth curves are derived from the Michaelis-Menten equation $y = ax/(b + x)$. The half satu-ration value (b) was 0.13 for the magnocellular-projecting ganglion cells and 1.74 for the parvocellular-projecting cells. Redrawn from Kaplan and Shapley (1986).

phenomenon of contrast induction seen in the patches of sinusoidal grating that are shown in Fig. 3.1. In Fig. 3.1 a-c, all three central patches have the same physical contrast when measured with a photometer. However, the perceived contrast of the respective patches depends strongly on the contrast of the surrounding gratings, so the central patch surrounded by the high contrast grating at the same spatial frequency (top middle) and orientation (middle left) as the central patch is judged to have a lower apparent contrast than the patch that is surrounded by the grating with the same contrast but a different spatial frequency (top left and right) or orientation (middle left or right). It should be noted that the overall mean level of illumination is the same across the whole of the figure, and is equal to the grey of the background. Not only is this effect seen for achromatic stimuli (Chubb et al., 1989) but it is also evident in chromatic gratings (Singer and D'Zmura, 1994, 1995).

Psychophysically, it has been demonstrated that some of the effect illustrated in Fig. 3.1 has a cortical locus because the adapting surround can be introduced through one eye while the central test patch is presented in the other eye; under these circumstances, the adaptation is still evident. The first place in the visual pathway that the inputs from the two eyes are mixed is at the level of striate cortex. The cortical locus is also supported by studies of effects of surrounds of different orientations on the apparent contrast of the central test patch (Fig. 3.1, middle row).

However, physiological studies have identified both retinal and cortical loci for contrast gain effects. The retinal mechanisms of contrast gain control are thought to be due to the network of lateral retinal connections. In the primate retina changes in ambient contrast result in a reduction in the gain of retinal M type ganglion cells to contrast (Benardete et al., 1992; Benardete and Kaplan, 1999) but there is little effect on P type ganglion cells (Benardete et al., 1992). In other words, for a given contrast level there is reduction in response of the M-cells if the ambient level of local contrast is high.

The retinal M-cells have a high contrast gain (Fig. 3.2). There is a relatively large increment in response in M-cells to small changes in contrast. The response maximum is attained at rather modest levels of contrast in retinal M-cells, typically 10-20% contrast (Kaplan and Shapley, 1986). At higher contrasts there is a region of response saturation where any further changes in stimulus contrast give the same, maximal, response (Fig. 3.2). In the chromatically opponent retinal P-cells, there is little evidence of response saturation, their response grows rather slowly as contrast changes and is often proportional to contrast (Fig. 3.2b).

The nonlinear retinal effects of contrast might contribute to some of the psychophysically observed effects of contrast normalization that appear to be rather nonspecific for stimulus dimensions and that are effectively monocular and achromatic. But the main effects of contrast normalization seem to be rather specific for stimulus spatial frequency, orientation, and chromaticity (Chubb et al., 1989; Solomon et al., 1993; Singer and D'Zmura, 1994, 1995). Thus, we are left with a considerable effect of local contrast that appears to be cortical in origin, the first place in the visual pathway in the primate where there is narrow spatial frequency tuning, orientation selectivity and binocularity. Furthermore, because retinal and LGN P-cells are not subject to contrast-gain effects, it appears reasonable to suggest that the interaction between chromatic and luminance signals that alter gain first appear at the level of the cortex.

The advantages of a system that shows adaptation, sometimes called normalization, is that there is a reduction in the effects of saturation. One of the properties that contrast adaptation confers on the visual representation is an invariance for spatial and orientation tuning. For example, in cat V1, the orientation tuning does not change at different levels of contrast (Ohzawa et al., 1985; Carandini et al., 1997). There is considerable support for the idea that the tuning of cortical neurones for spatial frequency and orientation is relatively invariant with contrast (Movshon et al., 1978; Ohzawa et al., 1985; Carandini et al., 1997). For a neurone that has a divisive process underlying normalization (Fig. 3.3a), we would expect the contrast-response functions for nonoptimal stimuli to be parallel to one another in log-log coordinates (Fig. 3.3b). The three curves shown in Fig. 3.3b correspond to the functions obtained at three off-peak orientations. Each of the functions is parallel to the next, showing the effects of normalization. The tuning of a hypothetical

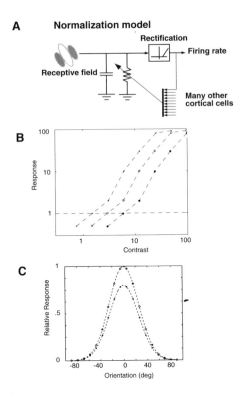

FIGURE 3.3. Normalization model (a) based on a linear receptive field passing through a thresholding rectification stage. The response (b) is dependent on the activity of a pool of local neurones that act as the normalization signal. One consequence of the normalization signal is that the contrast-response functions for off-optimal signals (open and filled circles) are shifted down relative to the optimal response (crosses). (c) shows a hypothetical pair of orientation tuning curves assuming a normalization of the response due to contrast.

V1 cell is shown in Fig. 3.3c. The response of the cell to a high contrast stimulus of the optimal spatial and temporal frequency for a range of orientations results in a characteristic orientation-tuning curve. The full bandwidth is about 40 deg., which is average for orientation tuned cortical cells (DeValois et al., 1982; Parker and Hawken, 1988). When the contrast is reduced then the response is scaled by a constant factor, in this case by 0.75. The orientation bandwidth of the cell remains constant, as does the peak of the tuning curve and the preferred orientation.

3.4   Beyond the Classical Receptive Field

Both neural models (Heeger, 1992) and psychophysically based models (Chubb et al., 1989; Singer and D'Zmura, 1995) of contrast normalization are based on a multiplicative stage. In the pure multiplicative process, the main effect is to move the contrast response function to the right with increasing normalization, as shown in Fig. 3.3. In normalization models, the area of visual space that gives rise to the normalization signal is not well-specified, except to say that it is local. Recent studies suggest that there can be quite distinct areas that are beyond the classical receptive field that can modulate the response of stimulation of the classical receptive field (Blakemore and Tobin, 1972; Maffei and Fiorentini, 1976; Nelson and Frost, 1978; Allman et al., 1985). A question that naturally arises is whether the area underlying the classical receptive field and the area from beyond the classical receptive field summate to produce the normalization signal?

Much of the current interest in surround effects hinges on a number of findings that suggest that influences from beyond the classical receptive field are more than alterations in contrast gain. They are purported to be very specific and able to aid in the representation of figure from ground (Zipser et al., 1996; Kapadia et al., 1999), to link object features (Kapadia et al., 1995; Polat et al., 1998), or in contextual segregation (Levitt and Lund, 1997). Other suggestions concerning modulation of the contrast gain concern the effects of attention, whereby attending to a target can specifically increase the response of the neurones that are signaling the presence of the target to the observer. The converse - that is, inattention - leads to a reduction in gain or suppression of the response (McAdams and Maunsell, 1999). These processes are thought to be specific and more "high level" in perceptual terms than the rather automatic gain control that is due purely to changes in local levels of contrast.

3.5   Area Summation and Contrast

In a preliminary study of receptive field spatial summation, we found that the spatial characteristics of cortical receptive fields could be modulated in specific ways. This suggests that the size of the summation region of cortical receptive fields is not an invariant feature of receptive-field organization. In the course of investigations of influences from beyond the classical receptive field, we were measuring the summation area of the classical receptive field so that we could position stimuli outside the classical receptive field. During these experiments, we made estimates of the summation area with stimuli of different contrasts and were struck by the consistent observation that the areal summation region appeared to be contrast-dependent. The summation area appeared to be considerably greater at low contrasts than at high contrasts (Sceniak et al., 1999). This is illustrated by the results of a typical area summation experiment shown in Fig. 3.4.

In this experiment, a circular patch of sinusoidal grating is located over the center of the classical receptive field. The orientation, spatial frequency, temporal frequency, and direction of the drifting grating are optimized for the neuron under study. Then, the response is determined for different circular patch sizes. Prior to this experiment, we measure the contrast response function for the cell so that we can choose contrasts that are on the quasi-linear portion of the contrast

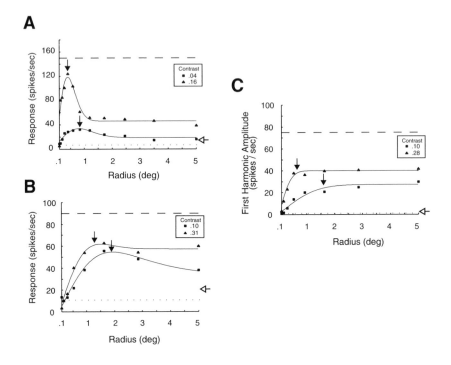

FIGURE 3.4. Spatial summation response profile for three V1 neurones. The visual stimulus was a circular patch of sine wave grating at the cell's preferred orientation, spatial frequency, and temporal frequency. The response magnitude is the mean rate (spikes/sec) for complex cells and the first harmonic amplitude for (spikes/sec) for simple cells. Low contrast responses are shown as filled squares and high contrast responses as filled triangles. (a) Summation profile for a representative complex cell from layer 4B. Low and high contrasts are 0.04 and 0.16 respectively. The excitatory space constant ratio (a_{low}/a_{high}) is 2.62. (b) Summation profile for a complex cell in layer 6 at low (0.10) and high (0.31) contrasts. The ratio a_{low}/a_{high} = 1.48. (c) Spatial summation curves for a simple cell in layer 6 at low (0.10) and high (0.28) contrast. The ratio a_{low}/a_{high} = 2.11. Contrast levels were chosen such that the dynamic range of the cell was covered while avoiding saturation at high contrast. The dotted horizontal line indicates the mean spontaneous firing rate. Open horizontal arrows show two standard deviations above the mean spontaneous firing rate. The dashed line shows each cell's saturating response. The vertical arrows indicate the radius of the peak response (maximum of the fitted curve), or if there were no clear maximum response, the radius of the response that first reached 95% of the maximum. Such summation radii were determined for low and high contrasts. The smooth curves are the best fits of the difference-of-Gaussians model described by DeAngelis et al. 1994 in Sceniak et al., (1999). Modified from Sceniak et al. (1999).

response function (see Fig. 3.2). For a low contrast we chose a point more than two standard deviations above the spontaneous response level on the contrast response function, whereas for a high contrast we chose points that are below 90% of the maximal response. Selection of these contrasts ensures that we do not confound our measure by using contrasts that produce saturating responses. In Fig. 3.4a the response to low contrast (4%) is shown by the squares while the response to the high contrast (16%) is shown by the triangles. The point on each curve that denotes maximal summation is shown by the solid vertical arrow. For the low-contrast area-summation curve, maximal summation is reached at about 0.9 deg., whereas for the high-contrast curve, the maximal summation is at about 0.35 deg, a ratio of 2.6. In the area-summation curve, there is a distinctive suppression of the response at areas greater than the optimal summation area. It was a natural question to ask whether the suppression with increasing area was also increasing with contrast, leading to an apparent change in the summation area. The alternative hypothesis is that there is an increase in the summation area of the classical receptive field.

To address this question, we turned to the use of a model of the receptive field that includes both an excitatory summation region, which can be thought of as the classical receptive field, and a suppressive region, which encompassed both the classical receptive field and the area beyond (DeAngelis et al., 1994). The components of the model are shown in Fig. 3.5. In the model there are center and surround Gaussian regions that summate as in the difference of Gaussians used to model the classical receptive field of retinal ganglion cells (Rodieck, 1965; Enroth-Cugell and Robson, 1966). It should be emphasized that the central Gaussian region includes all of the subregions of the classical receptive field of a typical cortical neuron, so that stimulation in the centre region would give the characteristic spatial frequency and orientation tuning of a cortical neuron. When we apply this model to the area summation curves that are shown in Fig. 3.4 we obtain estimates of the central space constant (a) and amplitude (K_e) and the surround space constant (b) and amplitude (K_i).

The distribution of the center space constants is shown in Fig. 3.6a for two different levels of contrast and the ratio is shown in Fig. 3.6d for a population of 85 neurones. There is a clear shift in the center space constant towards larger values at the lowest contrast. On average, there is a 2.3-fold increase in the summation radius at low contrast compared to high. The effect of contrast is not confined to the ends of the receptive field as it would be predicted if this were an inhibitory or suppressive mechanism due to end-stopping. For experiments in which the length of the stimulus was varied systematically while stimulus width was held constant at the optimal value found in the area-summation experiment, we found that there was an expansion of nearly three fold in length summation at the low contrast compared to the higher contrast (Figs. 3.6b, e). When the length was kept constant at the optimal value and the width of the stimulus varied, again there was a just greater than twofold expansion in the width dimension (Fig. 3.6c, f).

Because we have obtained the optimal parameters for the model applied to the area-summation curves, it is important to ask whether the change in summation area is uniquely due to the change in space constant of the central region or whether

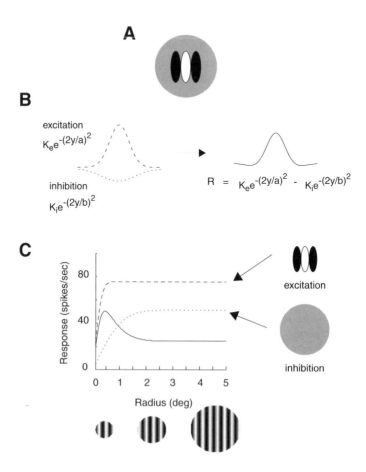

FIGURE 3.5. Difference-of-Gaussians model (DOG) for spatial summation. (a) Center-surround receptive field organization. The excitatory receptive field ON-OFF subunits are shown as white and black ellipses respectively. This is the classical receptive field whose extent is determined by the space constant of the center Gaussian. The gray region surrounding the subunits represents a suppressive surround. (b, c) The DOG model as used here assumes overlapping Gaussians for the excitatory subunits and the suppressive surround. The regions are combined linearly. The solid line represents the integrated area under the DOG model. The dotted line represents the inhibitory component and the dashed line the excitatory component, as illustrated by the diagrams pointing to each component.

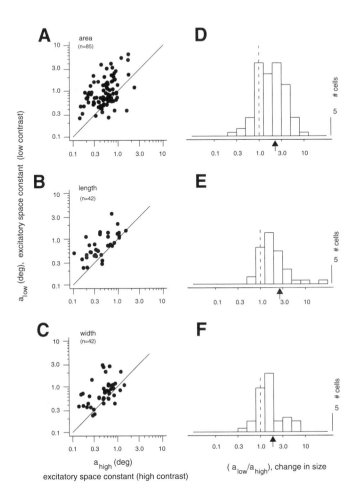

FIGURE 3.6. Dependence of excitatory spread on stimulus contrast. (a-c) Scatter plot of excitatory space-constant parameter at high versus. low contrast measured for circular patch summation, length summation, and width summation, respectively. (d-f) Histogram of the entire population, showing the ratio a_{low}/a_{high} of the excitatory spread at low to high contrast for each cell for area, length and width summations respectively. A value greater than unity indicates that the excitatory space constant is smaller when measured with high-contrast stimuli. The arrow indicates the population average. From Sceniak et al. (1999).

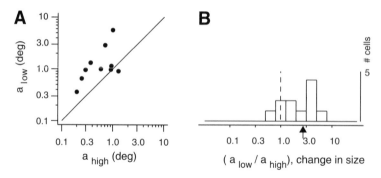

FIGURE 3.7. Cells that show no surround suppression at either contrast. (a) Scatter plot of the excitatory space constant at low versus high contrast for all cells $N = 11$) within the population that show no surround suppression. (b) Histogram of the ratio (a_{low}/a_{high}) for cells with no surround suppression. The arrow indicates the population mean (mean $= 2.7$).

a single space constant could be applied to the area summation curves and whether with a suitable adjustment in gain (K_e and K_i) could there be an adequate fit to the data. We addressed this question and, in some cases, a change in gain with a single space constant can fit the two area response curves adequately. In most cases we have studied, a change in center space constant was required to account for the results.

Further evidence to support the idea that there is a change in the excitatory space constant, representing the size of the classical receptive field, comes from area-summation curves such as the one shown in Fig. 3.4c. In neither case, the low nor the high contrast condition, is there any suppression at large areas. There is an increase in response with increasing stimulus area that reaches a plateau. This plateau is well below the saturation ceiling for this neuron,s shown by the dashed horizontal line in Fig. 3.4c at about 75 spikes/sec. Yet for this class of neuron, the area at which the response reaches the plateau for the low contrast condition is clearly greater than the area that is required for the response to plateau in the higher contrast condition. There is no evidence of a surrounding suppressive region and, in the model that we have applied, the only parameters that can change are those that affect the central summation region. To account for the greater areal summation at the lower contrast we need to increase the central space constant. Figure 3.7a shows the best fitting center space constants for the low and high contrast conditions for 11 neurones that show no surround inhibition. Points above the line of equality indicate neurones where the space constant obtained at low contrast was greater than the space constant obtained at high contrast. All the points are either on or above the line of equality, which suggests that there is an increase in space constant at low contrasts. This is emphasized in the histogram of the ratio of the excitatory space constants at low to high contrasts in Fig. 3.7b. The mean of the ratio, indicated by the solid arrow, is 2.7.

3.6   Temporal Tuning and Contrast

Contrast gain effects have been extensively documented in the retina of the cat (Shapely and Victor, 1978; 1981; Victor, 1987) and more recently in the primate retina (Benardete et al, 1992; Benardete and Kaplan, 1999). For cat X- and Y- cells, there appears to be a significant temporal-frequency-dependent effect of contrast. For cells in the primate magnocellular pathway, which has mainly linear cells with X-like spatial summation properties, there is a strong contrast gain effect at low temporal frequencies (Benardete et al, 1992, Benardete and Kaplan, 1999). This decrease in responsiveness of retinal M-cells is predominantly at low temporal frequencies, with little or no effect at high temporal frequencies. At first glance this might seem to be opposite to the effects reported in the cat X- and Y-cells, where "one effect of contrast is the relative enhancement of the amplitudes of responses to higher temporal frequencies at higher contrasts" (Shapley and Victor, 1981). The underlying effect is almost certainly the same for both primate M-cells and for cat X-cells; at low temporal frequencies there is a sharp attenuation of the responses with increasing contrast which leads to the "relative enhancement" of higher temporal frequencies. Any effects of contrast on the temporal transfer functions of V1 neurones that have a predominant M-cell input might be a combination of the effects that are generated in the retina and additional effects that are introduced at the level of V1.

The gain of retinal P-cells as a function of temporal frequency tends to be unaffected by contrast (Benardete et al., 1992). We obtained similar results for P-cells in the primate LGN (Fig. 3.8). It seems as if the input to V1 from the parvocellular LGN is relatively free of retinal contrast gain effects. Consequently any effects of contrast that are found for cells in the V1 that have a dominant P-cell input can most likely be attributed to a cortical locus. In general, neurones in the M-cell pathway show either no response to equiluminant chromatic modulation or very weak responses, and this residual response to equiluminant chromatic modulation is only evident when the chromatic contrast is high. Thus, it is most likely that the responses in the cortex to equiluminant red/green or yellow/blue stimuli are due to input to the cortex via the P-pathway. If one sees contrast dependent changes in cortical responses to such purely chromatic stimuli, it is reasonable to suppose such effects are entirely cortical.

3.7   Temporal Tuning and Contrast in V1

Figure 3.9a shows responses of a V1 cell to achromatic gratings of different contrasts. Unlike the effects in the LGN, the main effects in the cortex are an attenuation of the high temporal-frequency response at lower contrasts. If the responses are normalized, for contrast as in Fig. 3.9b, to show gain, then the clear attenuation of the high temporal frequency response at low contrast remains undiminished. Furthermore, there is little attenuation to low temporal frequencies; the neuron is

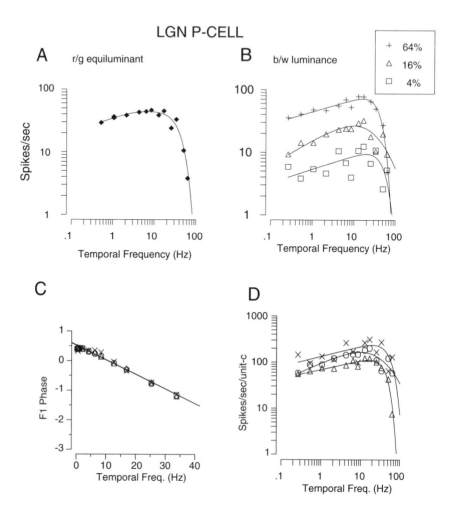

FIGURE 3.8. P-cell from the lateral geniculate nucleus of the macaque. (a) Shows the temporal frequency tuning to an isoluminant red/green chromatic grating at the optimal spatial frequency. (b) The temporal frequency tuning functions at different levels of contrast (4%, 16% and 64%) for a black/white achromatic grating. (c) Response phase of for the gain plot shown in (d). The response phase overlaps at all contrasts. The gain of the response, the values of response in (b) divided by contrast. The x's are 4%, squares 16% and triangles 64% contrast, corresponding to the squares, triangles and +'s in (b). It is clear that in the gain plot the responses are essentially overlapping for all three contrast levels.

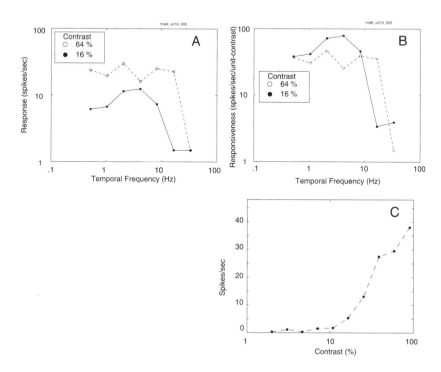

FIGURE 3.9. V1 simple cell temporal-frequency tuning at low and high contrasts. (a) The tuning at high contrasts shows an substantial difference between the low contrast and high contrast tuning, there is a reduction in the response to high temporal frequencies at low contrast and a shift in the peak of the tuning curve towards lower temporal frequencies. (b) The responses in A have been divided by the contrast of the grating to give units of spikes/sec/unit contrast or responsiveness. Even though the peak responsiveness at 16% contrast is higher than at 64% contrast the attenuation of the high temporal frequency responses is still evident. (c) Contrast response function for the simple cell, showing a response with a high contrast threshold and little saturation even at high contrasts, similar to the contrast response of retinal P-cells.

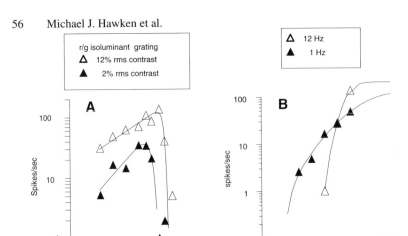

FIGURE 3.10. (a) Temporal frequency tuning for red/green isoluminant grating of an non-oriented V1 neuron from layer 6. The neuron showed low-pass tuning to spatial frequency. At highest level of cone contrast we could attain on our monitor (12%) the neuron is giving a very robust response to an optimal temporal frequency of about 12 Hz and then shows a very rapid attenuation of the response by about 20 Hz. At low contrast the response is already attenuated by 12 Hz. (b) The contrast response function for 1 Hz (filled triangles) and 12 Hz (open triangles) temporal modulation. There is a much higher gain and threshold for the 12 Hz temporal modulation.

essentially low-pass to temporal frequency. The low-pass nature of the tuning is evident at both low and high contrasts. It is likely that this cell gets a significant input from the parvocellular stream from the lateral geniculate nucleus. The contrast response function (Fig. 3.9c) shows that the cell has a relatively low contrast sensitivity (threshold is greater than 10%) and the gain is relatively linear up to the highest contrasts. This is similar to the contrast response function of retinal ganglion cells projecting to the parvocellular layers of the LGN (Fig. 3.2). In comparison with the temporal-frequency tuning of LGN P-cells at different contrasts (Fig. 3.8), the cortical effect of contrast tends to concentrate on the response to high temporal frequencies, which is not affected by contrast in LGN P-cells.

Further evidence that there is cortical filtering, especially at low contrast, for high temporal frequencies among some of the cortical population comes from a comparison of the tuning to chromatic gratings. The chromatic signal is carried to the cortex via the P-cell and I-cell pathways. The red/green opponent signal is in the P-cell pathway. At the level of the LGN, there is a robust response to equiluminant red/green gratings among the P-cells (Fig. 3.8a). This response is very similar to the response of the P-cell to achromatic gratings (Fig. 3.8b). In the cortex, there are two effects, one is an overall attenuation of the high-temporal-frequency response to chromatic stimuli and a further reduction as a function of contrast. This is illustrated in Fig. 3.10a. The response to a chromatically sensitive cell from layer 6 of V1 is shown to a relatively high chromatic contrast (12% RMS cone-contrast) and a low contrast (2% RMS-cone contrast). First, compared to the LGN P-cells the response to the high cone-contrast is reduced at high temporal frequencies (Fig. 3.10a, response to 12% contrast). Second, when the contrast is reduced there is a further reduction in the response to the higher temporal frequencies.

The pronounced contrast-dependent high-frequency attenuation in V1 is accompanied by characteristic changes in the contrast-response functions of V1 cells. For a V1 neuron that shows a significant attenuation of the high contrast response at lower contrasts, we want to compare the contrast-response functions at different temporal frequencies. If we chose two temporal frequencies, one close to the low-frequency cutoff of the tuning function and another close to the high-frequency cutoff, then when we determine the contrast response function for each temporal frequency we see quite different functions. At low temporal frequencies the contrast response function is shallow with a relatively low contrast threshold (Fig. 3.10b, filled triangles). At the higher temporal frequencies the contrast threshold is higher, and the slope of the contrast response function is much steeper than at low contrasts (Fig. 3.10b, unfilled triangles). A comparison of these functions to those of an LGN M-cell reveals quite different processes at work. The LGN cell contrast-response functions for low and high temporal frequencies are essentially parallel, reflecting the signature of a divisive process (Heeger, 1992), presumably the retinal contrast-gain control.

We fitted the temporal-frequency-tuning data on a population of V1 neurones with a difference of two exponential functions and obtained characteristic measures of the temporal tuning from the fitted functions (Hawken et al., 1996). To investigate the effect of contrast on the population, the two most informative descriptors of the tuning are the high temporal frequency cutoff and the peak of the tuning function. The distributions of high temporal-frequency cut-off for different conditions are shown in Fig. 3.11. The distribution of high temporal frequency cutoffs for high contrast black/white achromatic gratings ($> 64\%$) is shown by the histogram in Fig. 3.11d. For this condition the median is about 25 Hz. For the three other conditions – red/green equiluminance, S-cone isolating and low contrast ($8 - 16\%$) black/white gratings – there is a shift of about $1.5 - 2$ octaves of the median cut-off frequency. These differences in tuning for the chromatic stimuli are most likely to be a cortical effect because the LGN P-cells that provide the input to the cortex are giving responses to much higher temporal frequencies than are being passed through the cortical circuitry.

3.8   Discussion

One of the main functions of gain-control mechanisms is to maintain tuning of neurones to the features of the visual stimulus despite changes in local illumination and contrast. It is a mechanism that is used by the nervous system to maintain sensitivity with neurones of limited dynamic range in an environment where the amplitude of the physical stimulus can change by many orders of magnitude greater than the firing range of neurones. Our results suggest that there are major departures from models of the cortical gain-control mechanisms that envisage a static size of receptive field that is spatiotemporally invariant (Ohzawa et al., 1985; Carandini et al., 1997).

In the spatial domain, our results suggest that the envelope of spatial summation for V1 cells changes substantially with contrast (Sceniak et al., 1999). One of the adaptations to high contrast is a shrinkage of the summation region of the classical receptive field. The results from a large-scale model of V1 neurones suggest that results similar to those we found for areal summation could arise due to an increase in surround suppression at higher contrasts (Somers, 1998). We determined whether or not there was a correlation between the strength of the surround and contrast. In our results, there was no correlation (Sceniak et al, 1999). This supports our conclusions that there is a real change in the size of the summation region rather than an apparent change due to an increase in surround inhibition. One of the consequences of a reduced envelope of areal summation for the classical receptive field at high contrast is an increase in the localization sensitivity at the expense of spatial frequency tuning (Daugman, 1985). One speculation that we have made previously is that the increase in the summation area at low contrasts will help increase sensitivity to low contrast features of the image. Therefore there is an interplay between an increase in sensitivity and with decreasing localization at low contrast and decreased sensitivity with increased localization at high contrasts.

In the temporal domain, the tuning of V1 cells does not remain invariant with contrast. The results reported here are in agreement with the increased bandwidth of the temporal-tuning function in cat area 17 cells when stimulated with sum-of-sinusoids rather than single sinusoids in the temporal domain (Reid et al., 1992). Furthermore, examples of changes in temporal tuning with contrast have been noted in earlier studies (Hawken et al., 1992; Albrecht, 1995; Carandini et al., 1997). The cortical temporal filter seems to be ubiquitous, in that it affects all classes of cells. Of particular interest is are the chromatic responses because they are most likely carried by the P- and M-projecting cells of the retina. The P-cells show no alterations in their temporal tuning with contrast (Benardete et al., 1992). This makes the P-pathway an ideal system in which to study cortical gain control because, unlike the X and Y-cell pathways in the cat and the M-cell pathway in the monkey, there is no retinal gain control demonstrable in the P-cell pathway up to and including the LGN. Just as there are strong psychophysical correlates of static gain control mechanisms, there are also strong correlations between the cortical attenuation of high temporal frequencies and the attenuation of chromatic sensitivity at high temporal frequencies (Wisowaty, 1981) and the apparent slowing in speed at low contrasts (Thompson, 1982; Stone and Thompson, 1992, Hawken et al., 1994).

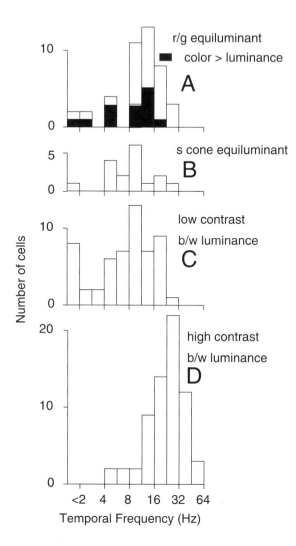

FIGURE 3.11. Distributions of the high-cutoff temporal frequency for different stimulus contrasts and chromaticities. The stimulus for each measurement was a drifting grating of the optimal spatial (frequency, orientation and direction. Temporal-frequency-tuning functions were obtained for each cell and fitted with a difference-of-exponentials function (Hawken et al., 1996). The response to chromatic gratings (a and b) and low contrast (8-16%) luminance gratings. (c) at high temporal frequencies is attenuated compared to high-contrast luminance gratings (d) The filled portions of (a) are for cells in which the response to the chromatic grating was greater than the response to a black/white (b/w) luminance grating.

Acknowledgments

This research was supported by NIH grants EY-08300 (MH) and EY-01472 (RS), by NSF, and the Sloan Foundation.

References

Albrecht, D. G. (1995). Visual cortex neurons in monkey and cat: effect of contrast on the spatial and temporal phase transfer functions. *Vis. Neurosci.*, 12:1191-1210.

Albrecht, D. G. and Geisler, W. S. (1991). Motion sensitivity and the contrast response function of simple cells in the visual cortex. *Vis. Neurosci.*, 7:531-546.

Allman, J., Miezin, F. and McGuinness, E. (1985). Stimulus specific responses from beyond the classical receptive field: neurophysiological mechanisms for local-global comparisons in visual neurons. *Ann. Rev. Neurosci.*, 8:407-430.

Benardete, E. A., Kaplan, E. and Knight, B. W. (1992). Contrast gain control in the primate retina: P cells are not X-like, some M cells are. *Vis. Neurosci.*, 8: 483-486.

Benardete, E.A. and Kaplan, E. (1999). The dynamics of primate M retinal ganglion cells. *Vis. Neurosci.*, 16:355-368.

Blakemore, C. and Tobin, A. E. (1972). Lateral inhibition between orientation detectors in the cat's visual cortex. *Exp. Brain Res.* 15:439-440.

Carandini, M., Heeger, D. J. and Movshon, J. A. (1997). Linearity and normalization in simple cells of the macaque primary visual cortex. *J. Neurosci.*, 17:8621-8644.

Chubb, C., Sperling, G. and Solomon, J. A. (1989). Texture interactions determine perceived contrast. *Proc. Nat. Acad. Sci.*, 86:9631-9635.

Craik, K. J. W. (1938). The effect of adaptation on differential brightness discrimination. *J. Physiol.*, 92:406-421.

DeAngelis, G. C., Freeman, R. D. and Ohzawa, I. (1994). Length and width tuning of neurons in the cat's primary visual cortex. *J. Neurophysiol.*, 71: 347-374.

DeValois, R. L., Yund, E. W., and Hepler, N. (1982). The orientation and direction selectivity of cells in macaque visual cortex. *Vis. Res.*, 22:531-544.

D'Zmura, M. and Singer, B. (1999). Contrast gain control. In K. R. Gegenfurtner and T. L. Sharpe (Eds.) *Color Vision: From Genes to Perception*, pp. 370-385, Cambridge: Cambridge University Press. Cambridge.

Daugman, J. D. (1985). Uncertainy relation for resolution in space, spatial frequency, and orientation optimized by two-dimensional visual cortical filters. *J. Opt. Soc. Am. A*, 2:1162-1169.

Enroth-Cugell, C. and Robson, J. G. (1966). The contrast sensitivity of retinal ganglion cells of the cat. *J. Physiol.*, 187:517-552.

Hawken, M. J., Gegenfurtner, K. R., and Tang, C. (1994). Contrast dependence of colour and luminance motion mechanisms in human viison. *Nature* 367: 268-270.

Hawken, M. J., Shapley, R. M. and Grosof, D. H. (1992). Temporal frequency tuning of neurons in macaque V1: effects of luminance contrast and chromaticity. *Invest. Ophthalmol. and Vis. Sci.*, 33(suppl.): 1313.

Hawken M. J., Shapley R. M. and Grosof D. H. (1996). Temporal-frequency selectivity in monkey visual cortex. *Vis. Neurosci.* 13:477-492.

Heeger, D. J. (1992). Normalization of cell responses in cat striate cortex. *Vis. Neurosci.* 9:181-197.

Hubel, D. H., Wiesel, T. N. (1962). Receptive fields, binocular interaction and functional architecture of cat's visual cortex. *J. Physiol.* 160:106-154.

Hubel, D. H., Wiesel, T. N. (1968). Receptive fields and functional architecture of monkey striate cortex. *J. Physiol.*, 195:215-245.

Kapadia, M. K., Ito, M., Gilbert, C. D. and Westheimer, G. (1995). Improvement in visual sensitivity by changes in local context: parallel studies in human observers and in V1 of alert monkeys. *Neuron.*, 15:843-856.

Kapadia, M. K., Westheimer, G. and Gilbert, C. D. (1999). Dynamics of spatial summation in primary visual cortex of alert monkeys. *Proc. Nat. Acad. Sci.*, 96:12073-12078.

Kaplan, E. and Shapley, R. M. (1986). The primate retina contains two types of ganglion cells, with high and low contrast sensitivity. *Proc. Nat. Acad. Sci., USA*, 83:2755-2757.

Levitt, J. B. and Lund, J. S. (1997). Contrast dependence of contextual effects in primate visual cortex. *Nature*, 387:73-76.

McAdams, C. J. and Maunsell, J. H. (1999). Effects of attention on orientation tuning functions of single neurons in macaque cortical area V4. *J. Neurosci.*, 19:431-441.

Maffei, L. and Fiorentini, A. (1976). The unresponsive regions of visual cortical receptive fields. *Vis. Res.*, 16:1131-1139.

Movshon, J. A., Thompson, I. D. and Tolhurst, D. J. (1978). Spatial summation in the receptive fields of simple cells in the cat's striate cortex. *J. Physiol.*, 283:53-77.

Nelson, J. I. and Frost, B. J. (1978). Orientation-selective inhibition from beyond the classic receptive field. *Brain Res.*, 139:359-365.

Ohzawa, I., Sclar, G. and Freeman, R. D. (1985). Contrast gain control in the cat's visual system. *J. Neurophysiol.*, 54:651-667.

Parker, A. J., Hawken, M. J. (1988). Two-dimensional structure of receptive fields in monkey striate cortex. *J. Opt. Soc. Am. A*, 5:598-605.

Polat, U., Mizobe, K., Pettet, M. W., Kasamatsu, T. and Norcia, A. M. (1998). Collinear stimuli regulate visual responses depending on cell's contrast threshold. *Nature*, 391:580-584.

Reid, C. R., Victor, J. D. and Shapley, R. M. (1992). Broadband temporal stimuli decrease the integration time of neurons in cat striate cortex. *Vis. Neurosci.*, 9:39-45.

Rodieck, R. W. (1965). Quantitative analysis of cat retinal ganglion cell response to visual stimuli. *Vis. Res.* 5:583-601.

Sceniak, M. P., Ringach, D. L., Hawken, M. J. and Shapley, R. M. (1999). Contrast's effect on spatial summation by macaque V1 neurons. *Nature Neurosci.*, 2:733-739.

Shapley, R. M. and Victor, J. D. (1978). The effect of contrast on the transfer properties of cat retinal ganglion cells. *J. Physiol.*, 285:275-298.

Shapley, R. M. and Enroth-Cugell, C. (1983). Visual adaptation and retinal gain controls. *Progress in Ret. Res.*, 3:263-346.

Shapley, R. M. and Victor, J. D. (1981). How the contrast gain control modifies the frequency responses of cat retinal ganglion cells. *J. Physiol.*, 318:161-179

Singer, B. and D'Zmura, M. (1994). Color contrast induction. *Vis. Res.*, 34:3111-3126.

Singer, B. and D'Zmura, M. (1995). Contrast gain control: a bilinear model for chromatic selectivity. *J. Opt. Soc. Am. A*, 12:667-685.

Solomon, J. A., Sperling, G. and Chubb, C. (1993). The lateral inhibition of perceived contrast is indifferent to on-center/off-center segregation, but specific to orientation. *Vis. Res.*, 33:2671-2683.

Somers, D. C. (1998). A local circuit approach to understanding integration of long-range inputs in primary visual cortex. *Cereb. Cort.*, 8:204-217.

Stone, L. S. and Thompson, P. (1992). Human speed perception is contrast dependent. *Vis. Res.*, 32:1535-1549.

Thompson, P. (1982). Perceived rate of movement depends on contrast. *Vis. Res.*, 22: 377-380.

Victor, J. D. (1987). The dynamics of the cat retinal X-cell center. *J. Physiol.*, 386:219-246.

Wisowaty, J. J. (1981). Estimates of the temporal response characteristics of the chromatic pathways. *J. Opt. Soc. Am.*, 71:970-977.

Zipser, K., Lamme, V. A. and Schiller, P. H. (1996). Contextual modulation in primary visual cortex. *J. Neurosci.*, 16:7376-7389.

4

Global Processes in Form Vision and Their Relationship to Spatial Attention

Frances Wilkinson
Hugh R. Wilson

4.1 Introduction to the Ventral Visual Pathway

One of the major advances in neuroscience over the past 40 years has been the anatomical unravelling of the primate visual pathway, particularly its cortical components. From the vantage point of the year 2000, it is difficult to realize how little was known of the territory beyond the striate cortex (V1) when Hubel and Wiesel began their single unit exploration of primate V1 in 1968 (Hubel and Wiesel, 1968), and Gross described object specific neurons in inferotemporal cortex in 1969 (Gross, Bender, and Rocha-Miranda, 1969). This despite the fact that fully half of the macaque neocortex is in some way involved in visual function. However, by 1991 over two decades of intensive anatomical and electrophysiological investigation by many investigators culminated in the publication by Felleman and Van Essen of a comprehensive representation of the parallel and hierarchical relationships among the more than 30 cortical areas with visual functions in the macaque brain (Felleman and Van Essen, 1991).

Until the mid-1980's, parallels between macaque and human visual organization beyond V1 were largely speculative, based primarily on comparisons between primate electrophysiology and lesion data and neuropsychological evidence from humans sustaining damage to cortical areas due to strokes, tumors or gunshot wounds. However, the advent of functional imaging methodologies in the 1980's has opened the way for closer examination of the cortical visual areas in the intact human brain. An excellent summary of the current status of the comparison between human and primate may be found in Van Essen and Drury (1997).

Despite these impressive anatomical gains, the functional roles of the majority of these multiple visual areas remain largely unknown. In this chapter, we focus on a subset of areas within the ventral visual stream (Ungerleider and Mishkin, 1982), those areas lying between V1 and the inferotemporal cortex. Based on both lesion data and single unit electrophysiology, it is clear that IT and the areas providing its inputs form the critical substrate of object recognition, leading to the labelling of the ventral stream as the "what" pathway. The raw material for this analysis, the encoding of the visual image in V1, is relatively well understood. Banks of orientation- and size- (or spatial frequency-) tuned filters extract local luminance transitions at multiple spatial scales (DeValois, Albrecht, and Thorell, 1982). In separate interdigitated modules (the "blobs"), chromatic contrast information is encoded over the same local regions by units that appear not to be orientation sensitive (Livingston and Hubel, 1984). In contrast, the cortical territory between V1 and IT remains poorly understood. It is widely assumed that the loosely hierarchical set of intervening regions (V2, VP, V4, TEO) must involve the sequential extraction of increasingly complex aspects of visual objects. However, due in large part to a dearth of strong models for shape analysis and object recognition, relatively little progress has been made either by single unit electrophysiology or by functional imaging in unravelling this process.

4.2   Components of Intermediate Form Analysis

Beginning with the local luminance and chromatic code in V1, one might anticipate several transformations as one moves up a form-analyzing hierarchy.

1. Contour orientation should be extracted as an emerging property of discontinuities in other visual properties such as stereodisparity, relative motion, and textural change, providing additional information about object boundaries and hence shape. The observation of neurons showing orientation tuning to illusory and texture boundaries in V2 (von der Heydt, Peterhans, and Baumgartner, 1984; von der Heydt and Peterhans, 1989) would be an example of such a process.

2. Information about local orientation might be expected to be pooled configurally across space in order to extract more complex shapes from the image.

3. Gradient information from any of the visual dimensions mentioned earlier (luminance, texture, disparity, motion) must be integrated to extract information about the 3-D surface structure of objects within the image.

4. Information must be combined across spatial scales. For example, common orientation information at several scales is indicative of a sharp edge, whereas information localized only at low spatial frequencies might indicate shading, and orientation signals limited to high spatial frequencies are indicative of texture. Objects defined predominantly by the configuration of their internal features (faces being a prime example here) require assessment of interscale spatial relationships.

5. As objects are dissociated from the background through such characteristics as continuous boundary (2 above) and continuous or smoothly varying

surface properties (3 above), one would anticipate increasingly strong suppression of ground by figure. This could be implemented through feedback loops such that structures identified at intermediate or high levels of the pathway would enhance signals associated with "figure" at the expense of "ground" through selective feedback facilitation (Tsotsos, 1990). Reciprocal connections are a ubiquitous feature of the cortical visual pathways (Felleman and Van Essen, 1991), so there is a clear anatomical basis for such a functional mechanism. This sort of downward-flowing facilitation could also reflect voluntary selective-attention mechanisms.

The focus of recent work in our laboratory has been on the second of these five mechanisms – the configural combination of orientation information. Our work was inspired by electrophysiological findings in primate area V4, so it is to that region that we now turn our attention.

4.3   Changing Views of V4

A pivotal area within the primate ventral stream is V4. In the macaque, V4 receives feedforward projections from all earlier cortical areas (Felleman and Van Essen, 1991). For example, both colour and orientation coded modules of V2 send projections to V4 (Felleman, Xiao, and McClendon, 1997). Moreover, V4 receives lateral projections from area MT in the dorsal stream (Maunsell and Van Essen, 1983), which might provide a route for boundary and gradient information extracted from motion signals to be incorporated into shape analysis. V4 is also the target of descending inputs from higher levels of the ventral pathway. Since its discovery by Zeki (1971), at least three roles have been ascribed to this cortical area: involvement in colour, selective attention, and form analysis. These are not, of course, mutually exclusive. The particular role of V4 in analysis of chromatic inputs is controversial, and will not be considered further here. Primate electrophysiological studies have demonstrated that the receptive field characteristics of V4 neurons are modulated by selective attention (Connor, Preddie, Gallant, and Van Essen, 1997; Luck, Chelazzi, Hillyard, and Desimone, 1997; Moran and Desimone, 1985; Motter, 1993), a phenomenon that also occurs at several other levels of the system. To date, studies of V4 selective attention have concentrated on very simple stimuli - too simple, we will argue, to fully expose the complexity of V4 function. Nevertheless, the evidence for attentional modulation is strong, and later in this chapter we will return to this subject to discuss evidence from our own work on competitive attentional networks at this level of the system. First, however, we will turn to the role of V4 in intermediate form vision. Is there evidence that V4 contributes to any of the processes outlined earlier in this chapter?

Unfortunately, few electrophysiological studies of V4 have used stimuli more complex than those optimal for V1 activation. However, when presented with more complex stimuli, V4 neurons provide tantalizing suggestions of involvement in the sorts of processes outlined earlier. Our interest was particularly captured by two reports by Gallant and Van Essen (Gallant, Braun, and Van Essen, 1993; Gallant, Connor, Rakshit, Lewis, and Van Essen, 1996) in which grating patterns, which they termed "non-Cartesian gratings", proved to be powerful stimuli for V4 neurons. These patterns consisted of concentric circles, radial lines, spirals, and hyperbolic gratings. All of these patterns require the pooling of a variety of orientations over specific spatial configurations - the second of the "intermediate form processes" outlined above. A very basic feature of visual anatomy - increasing receptive field size at a given retinal eccentricity as one ascends the ventral stream - dictates that this sort of pooling must occur beyond V1. Although this need not occur in V4, we will argue later in this chapter that V4 receptive-field sizes are very appropriate to this task.

The notion of hardwired configural organizations is not universally accepted. It immediately raises the issue of how many organizations are present. Could there be enough different configurations distributed across the visual field to support shape discrimination? Is there an appropriate basis set that would provide the building blocks of all possible shapes? We do not have a definitive answer to this issue, but would propose that there are doubtless several classes of such configural units, of which we have concentrated on characterizing two: circular or concentric units, and radial or cross-shaped units. Our emphasis here will be on the former, and we will suggest that even this single simple organization when viewed in a distributed network context provides a foundation for the analysis of a wide range of biologically significant shapes. Among biologically important stimuli, we focus on one in particular - the face. It has been repeatedly suggested that faces are so biologically "special" that the brain may have evolved separate mechanisms dedicated to face perception (Meadows, 1974; Moscovitch, Winocur, and Behrmann, 1997). Whether this turns out to be true or not, the capacity to efficiently handle face recognition will be one of the most critical tests of any model of primate form perception because the face as a primary route of primate social communication must have exerted enormous selective pressure on the evolution of form analytic mechanisms.

4.4   Evidence for Global Orientation Pooling in Human Vision

Motivated by the physiology in Gallant et al. (1993; 1996), we have recently pursued two independent lines of research both of which have yielded evidence of circular or concentric pooling of orientation in human vision (Wilson, Wilkinson, and Asaad, 1997; Wilkinson, Wilson, and Habak, 1998; Wilson and Wilkinson, 1998). First, in Glass patterns (Glass, 1969; Glass and Pérez, 1973) (Fig. 4.1a), global configuration is conveyed by the layout of dot pairs, which themselves give rise to local orientation signals. In psychophysical studies, we have shown that human subjects are more sensitive to concentric organization than to translational (vertical or horizontal) organization in Glass patterns when tested in a decreasing

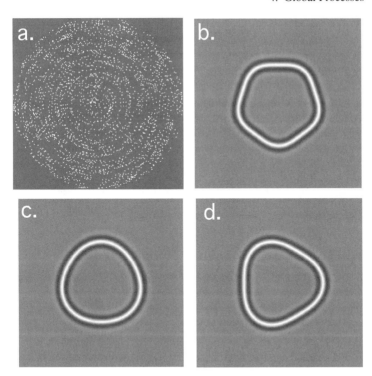

FIGURE 4.1. a) Concentric Glass pattern. Dot pairs of equal separation are aligned tangent to circles centered at the center of the pattern. b) Radial frequency pattern of 5 cycles (RF5). The frequency is the number of cycles of sinusoidal modulation of the radius of the pattern. c and d) RF3 patterns of differing amplitude (depth of modulation) and phase (angle of rotation).

signal-to-noise paradigm. Radial and hyperbolic organizations yielded intermediate thresholds. Second, by examining thresholds when signal-carrying sectors were separated by noise-only sectors, we obtained convincing evidence of linear summation in pooling of concentrically arrayed orientation signals. Using similar experimental approaches, we also found evidence for radial organization which, although weaker than the concentric structures, also involved summation of radially arrayed orientation signals (Wilson and Wilkinson, 1998).

    In an independent line of inquiry (Wilkinson et al., 1998), we have measured thresholds for detecting distortions of circles using a stimulus set we have called radial frequency (RF) patterns (see Fig. 4.1b-d). The frequency refers to the number of cycles of sinusoidal modulation of the radius introduced to our circular contour. Thresholds (amplitude of the sinusoidal variation) for detecting these distortions were found to be in the hyperacuity range across a wide range of radial frequencies, and detailed analysis reveals that these results cannot be explained by local orientation or curvature thresholds. Instead, we must appeal again to a global process of information combination. A recent report from Hess and collaborators (Hess, Wang, and Dakin, 1999) corroborates this conclusion. An important aspect of our radial frequency threshold data is that, expressed as Weber fractions (minimum detectable % change in radius), we found comparable thresholds across a wide range of stimulus radii and spatial frequencies. One implication of this finding is that the shape of a given RF pattern is the same at threshold independent of the specific size and spatial frequency structure of the pattern – an important instance of shape constancy (Wilkinson et al., 1998). Further experiments in our labs using a number of paradigms, have pointed to the lower radial frequencies (2–6) as being particularly important as possible contributors to shape discrimination, as they are all easily and immediately discriminable from one another.

4.5   Neural Model for Configural Units

Our psychophysical data from both Glass-pattern and radial-frequency studies point to the existence of configural units for both concentric and radial structure in the human visual pathway. We have developed a computational model of the underlying organization of these units using V1 simple cells as the first-level inputs to the model (Wilson et al., 1997). The basic model is outlined in Fig. 4.2. As illustrated here, a contrast-gain control is applied to the outputs of the simple cell filtering, followed by full-wave rectification. A second oriented filtering stage follows these operations and is presumed from the work of von der Heydt and colleagues (von der Heydt and Peterhans, 1989; von der Heydt, Peterhans, and Baumgartner, 1984) to reflect processing in V2. This filter-rectify-filter theme, which is common in models of both texture and second order motion perception (see Wilson, 1999a), generates a variety of complex receptive fields with end-stopping or suppressive surrounds. The key to generating such receptive fields is shown in Fig. 4.2: the second-stage filter must have a receptive-field central summation region approximately equal in width to the diameter of the first-stage filters feeding into it. For plausible oriented receptive-field filters, this requires that the second-stage filter be approximately three times the linear dimensions of the simple cells in V1. Finally, the third stage of our model involves configural pooling of orientation information around circular contours (Fig. 4.2a). We have shown only four component second stage inputs to the configural unit; however second stage units at 12 orientations are included in the model. By rotating the orientation of the second stage inputs to the configural stage, a similar mechanism can also give rise to units sensitive to radial organization (Fig. 4.2b), in this case a cross shape. We tentatively call these V4 configural units on the basis of their similarity in stimulus specificity to the neurons recorded in primate V4 by Gallant et al. (1993). It is important to emphasize that this same model schema can readily be adapted to produce many other V4 configural units optimized for other types of image structure, although psychophysical evidence only exists for concentric,

X-shaped, and cross-shaped configurations thus far (Wilson and Wilkinson, 1998).

An important feature of our full model is the incorporation of contrast gain controls between the first- and second-stage filters, an operation likely to occur in V1 (Heeger, 1992). One consequence of this is that the configural unit responses are independent of stimulus contrast over a substantial range (as was shown to be the case for RF discrimination thresholds – Wilkinson et al., 1998). This is illustrated in Fig. 4.3, where model responses to circles at 25% contrast are only slightly lower than responses to 100% contrast circles. As a result of this contrast gain control, the model response increases almost linearly with increasing circle diameter over the range from about 0.33° to 0.9°. The model response linearity results from summation at the V4 stage of activity generated around the circle circumference. This particular simulation was based on model V1 filters tuned to 16 cpd, but additional model units tuned to lower spatial frequencies would obviously extend this range to encode circle diameter over a much wider range.

Although a single concentric unit responds optimally to a circular pattern, these units also respond well to ellipsoidal shapes. Local networks of such units could act together to provide a powerful means of encoding a wide range of elliptical and quasicircular shapes, such as the radial-frequency patterns introduced earlier (Wilkinson et al., 1998). As will be discussed later, complex pseudoelliptical shapes such as human heads, can be described as a sum of RF components, and the encoding of such shapes could be distributed within a local network of concentric units.

4.6   Configural Units and Receptive Field Size

The argument that the component stages of our hierarchical model might lie in V1, V2, and V4 receives strong support from existing data on receptive-field sizes in these topographic representations of the visual field. It has long been recognized that receptive field size at a given retinal eccentricity increases as one moves up the ventral pathway; not only is there a qualitative match between this change and the requirements of the model, but also a quantitative fit. Typical receptive field diameter in macaque V1 fovea averages about 0.25° (Dow, Snyder, Vautin, and Bauer, 1981; Van Essen, Newsome, and Maunsell, 1984). This increases to a diameter of 2°-3° in the V4 foveal representation (Desimone and Schein, 1987; Gattass, Sousa, and Gross, 1988; Boussaoud, Desimone, and Ungerleider, 1991), and to a diameter of 6°-10° for foveal receptive fields in TEO (Boussaoud et al., 1991). V2 has also been shown to have larger receptive fields than V1 at all eccentricities, being around three times larger near the fovea (Gattass, Gross, and Sandell, 1981). Finally, receptive fields in TE, the highest level in the form-vision pathway, have been found to range from 25° to 40° in diameter and typically include the fovea (Gross, 1992). Putting these figures together, it can be concluded that mean foveal receptive field diameter increases by a factor of 2.5-3.0 at each higher level

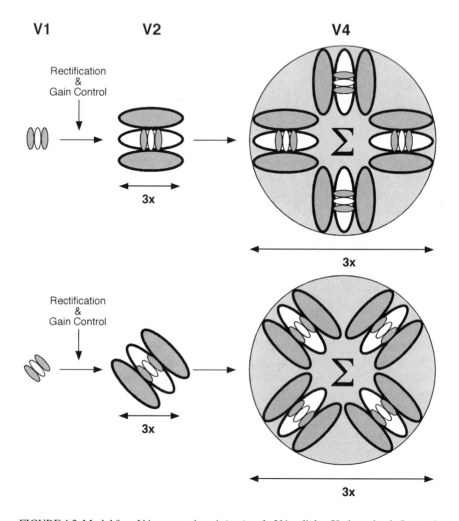

FIGURE 4.2. Model for a V4 concentric unit (top) and a V4 radial or X-shaped unit (bottom). The first stage incorporates model V1 simple cells at 12 different orientations (only two are shown). This is followed by contrast-gain control and rectification. Second stage oriented filtering is hypothesized to occur in V2 and must incorporate filters with linear dimensions about 3× those of their V1 inputs to extract appropriate information. The final model stage in V4 incorporates linear summation of concentrically arranged (top) or radially arranged (bottom) V2 inputs, thus producing units sensitive to global stimulus configurations. As the diagram illustrates, this final stage must again be about 3× larger in diameter than its inputs to extract configural information. Thus, this V4 configural model provides an explanation for the mean receptive-field size increase observed from primate V1 to V2 to V4.

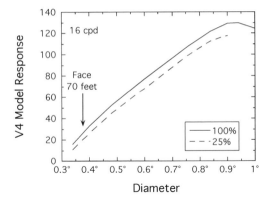

FIGURE 4.3. Response of model V4 concentric units to circular stimulus contours. In every case the maximum model response occurred at the center of the stimulus. The model contrast gain control renders the response almost independent of contrast over a considerable range, as is illustrated by responses to 100% (solid line) and 25% (dashed line) contrast circles. As a result of this contrast independence, the model response encodes stimulus diameter almost linearly over a range from 0.33° to 0.9°. This model incorporated V1 units tuned to 16 cpd and can encode face shape up to a distance of about 70 feet. Lower-spatial-frequency versions of the model will encode progressively larger stimulus diameters.

of the form vision hierarchy. This factor may be used to make estimates of mean receptive field diameter in the presumptively homologous areas of the human form vision system. As the peak of the foveal contrast sensitivity function is about 4.0 cycles/degree, a mean human V1 foveal receptive field can be inferred to be about 1.5 cycles of this frequency, or 0.375° in diameter. Assuming a factor of 3.0 for the diameter increase between areas, mean human foveal receptive field diameters may be inferred to be approximately 1.13° in V2, 3.38° in V4, 10° in TEO, and 30° in TE. These diameters agree rather well with the primate physiological literature cited earlier. Given that the range of peak spatial frequencies is similar in V1 and V4 (Desimone and Schein, 1987), this would still permit the smallest V4 configural units to encode areas of about 0.33° diameter (see Fig. 4.3).

4.7   Evidence Pointing to Configural Units in V4 in the Human Brain

In an investigation of the human form pathway using functional magnetic resonance imaging (fMRI), we have found strong evidence in support of V4 as a locus for these configural units. Whereas in V1, fMRI activation is equally strong in gratings formed of concentric circles, radial lines, and parallel lines (i.e., conventional sinusoidal gratings), in V4, the parallel gratings produce much weaker activation than the two other grating classes (James et al., 1999). This is precisely what would be predicted if orientation information were pooled in V4 in concentric and radial arrays but not along a translational axis – as indicated by our psychophysical results with Glass patterns.

Strong converging evidence has recently been provided by an electrophysiological recording study in human subjects (Allison, Puce, Spencer, and McCarthy, 1999). Using surface electrode arrays implanted prior to epilepsy surgery, visual responses were recorded from the inferior surface of the occipitotemporal region to a wide range of visual stimuli, including the Cartesian and non-Cartesian gratings used by Gallant et al. (1996), as well as faces, cars, butterflies and other objects. Early-event-related potentials associated with areas V1 and V2 (P100) showed sensitivity to the overall luminance, contrast, and size but not the configural properties of stimuli. However, electrodes placed somewhat more anteriorly on the inferior surface of the brain showed substantially greater responses to non-Cartesian (concentric circles, radial gratings, spirals, and hyperbolic gratings) than to conventional sinusoidal gratings at approximately 180 msec latency (N180). Still further anterior in the fusiform gyrus, an even later component of the evoked response (P300) showed strong selectivity for faces. In interpreting their data, Allison et al. (1999) suggest that the strong N180 response seen to non-Cartesian gratings probably arises from an intermediate stage of human form processing analogous to primate V4. The location of their recording electrodes strongly suggests that this may indeed be the same region we have studied using fMRI.

4.8   Application of V4 Model Units to Faces

What relevance might these hypothesized concentric configural units have to face perception? Two complementary roles suggest themselves. Within the form vision literature there is a distinction between boundary-based models and object-centered models (Burbeck and Pizer, 1995; Kimia, Tannenbaum, and Zucker, 1995). The problem with a purely boundary based description is the difficulty of interrelating parts both within and across scales. Our V4 model units incorporate elements of both boundary and object-centered models. While coding information about the bounding contour of a shape, these units also represent the centroid of the shape - that is, they are localized at the circle's centre. In the case of faces, image processing with an array of such units localizes the centre of the head by the unit of greatest activity. This can be used as an anchor for defining configural relations across spatial scales. In the case of the face, this would, for example, entail the location of the internal features such as the eyes. The power of concentric units for picking out a face in a complex scene is illustrated in Fig. 4.4. When this image was convolved with V4 model concentric units, the strongest activation occurred at the centre of the head (black dot), and the model activity at this point encoded the radius (see Fig. 4.3) shown by the vertical arrow. Clearly, therefore, the V4 model is capable of detecting head shapes even in transparent scenes.

Secondly, we have found that head shape can be well fit by summing lower-order

FIGURE 4.4. Model V4 concentric-unit response to a transparent face-house combination similar to those employed in studies of object-based attention by O'Craven et al. (1999). Despite the superimposed house, V4 concentric units only responded to the head, with the maximum response occurring at the point indicated by the solid circle. Based upon Fig. 4.3, the maximum response encoded the mean stimulus radius indicated by the arrow, and a population of adjacent unit responses indicated that the stimulus was ellipsoidal with a vertical major axis. This example demonstrates that the V4 model is quite flexible in localizing and measuring ellipsoidal shapes such as heads in complex scenes.

radial-frequency patterns (RF 1-5) in appropriate relative amplitudes and phases (see Fig. 4.5). In psychophysical studies, we have shown (Wilson, Wilkinson, Lin, and Castillo, 2000b) that head shape provides an important cue to head direction – complementary to and equally strong as internal facial features. Because gaze direction – an extremely important social cue – involves the combination of head direction and eye position relative to the head, the ability to reliably compute head direction is of great biological advantage.

Clearly the role of the concentric and radial units we have proposed is not limited to face perception. Many biological objects have shapes whose wholes (e.g. fruit) or parts (heads, trunks of animals) can be described by relatively simple convex closed curves. Similarly, cross- and X-shaped units may encode facial feature configurations but are also valuable in encoding a wide range of other object features. Thus, these configural form-extracting units lying at an intermediate level of the form hierarchy may provide critical building blocks for the encoding of complex, biologically meaningful objects.

4.9   Selective Attention

A number of physiological studies on alert primates have found evidence for attentional modulation of single neuron responses in V4 (Motter, 1993, 1994; Connor et al., 1997; Luck et al., 1997; McAdams and Maunsell, 1999; De Weerd et al.,1999). These studies also demonstrate that attentional effects are significantly

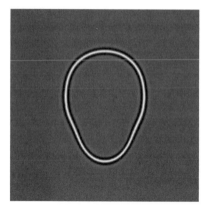

FIGURE 4.5. Simulated head outline synthesized through addition of appropriate amplitude and phase of RF2, RF3, RF4, and RF5 patterns of equal radius.

more pronounced in V4 than in either V1 or V2. In a survey of much of this literature, Desimone and Duncan (1995) concluded that attentional effects could be explained as a biasing of competitive neural networks. Recent physiological work has corroborated the hypothesis that neural competition forms the basis for selective attention (De Weerd et al., 1999; Reynolds, Chelazzi, and Desimone, 1999).

These data suggest the possibility that V4 configural units might play a significant role in selective attention. We have found support for this possibility in an analysis of a striking visual illusion (Wilson, Krupa, and Wilkinson, 2000a). Examination of Fig. 4.6 will generate a percept of illusory oscillating circles throughout the pattern, which is known as a Marroquin (1976) pattern. This pattern was produced by superimposing three copies of a square dot lattice: the original plus two copies rotated by $\pm 60^{\circ}$ about the pattern center. Although this pattern had been in the literature for 24 years, the only attempt at an explanation of the visual dynamics had been the suggestion that it somehow tapped dynamic grouping properties of the visual system (Stevens and Brookes, 1987). It was natural, therefore, to wonder whether processing the Marroquin pattern by model V4 concentric units might elucidate the underlying dynamics.

Psychophysical measurements were first made in order to quantify the phenomenon (Wilson et al., 2000a). Using methods similar to those employed to study binocular rivalry, the mean duration of circle visibility for two subjects was found to be about 2.8 sec with a standard deviation of about the same magnitude. In addition, the visibility intervals were well fit by a Gamma distribution, $(\lambda t)^n e^{-e\lambda t}$, with $\lambda = 1.7$ and $n = 1.8$ on average for these two subjects. This is quite similar to the visibility distribution for binocular rivalry (Fox and Herrmann, 1967; Blake, 1989). Additionally, the visibility of illusory circles was found to vary dramatically but reproducibly when measurements were made centered on different points in the pattern.

As a first step towards explaining the Marroquin illusion, the Marroquin pattern

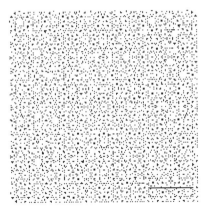

FIGURE 4.6. Marroquin pattern produced by superimposing three periodic dot grids oriented at $\pm 60°$ relative to one another. At a normal reading distance, illusory circles with the diameter indicated by the horizontal line oscillate about different points in the pattern.

was processed by the V4 concentric unit model. In agreement with the human psychophysical data, maximum model V4 responses occurred at points where circle visibility was maximal, whereas minimal model responses occurred at points where circles were seldom seen (Wilson et al., 2000a). To account for the oscillations triggered by the illusion, two further components were necessary in the model. The first was revealed by masking experiments showing that superimposed luminance circles dramatically reduced illusory-circle visibility durations if the masking circles were within a region surrounding the illusory-circle locus. Outside this region the masking rapidly fell to zero. There is thus experimental evidence for lateral inhibitory interactions between V4 concentric units, and this inhibition is spatially restricted.

To develop a model to explain illusory circle oscillations, lateral inhibitory connections were introduced among model V4 units, with the extent of this inhibition determined by the masking experiments. This produced what may be termed a spatially regional competitive or "winner take all" network. If such a network is p-resented with a pattern of appropriate stimuli, circles in this case, the most strongly stimulated unit in each region defined by the range of lateral inhibition will win the competition and suppress all surrounding neural activity within the region. For such a network to generate oscillations, however, there must also be some form of self-adaptation such that the response of each regional winner eventually decays to the point where another winner can emerge within the region. Such self-adaptation or negative feedback is known to be a general requirement for the emergence of oscillations in any dynamical system (Wilson, 1999b). A natural candidate for such self-adaptation is spike-frequency adaptation, which is ubiquitous in human and mammalian excitatory (but not inhibitory) neocortical neurons (Connors and Gutnick, 1990; Lorenzon and Foehring, 1992). Spike-frequency adaptation is produced by a calcium mediated potassium current that slowly hyperpolarizes excitatory cortical neurons only after they begin firing. In consequence the firing rate of excitatory cortical cells typically drops to about 33% of the initial transient rate within several hundred milliseconds.

With the inclusion of both regional competitive winner take all inhibition and spike-frequency adaptation, the V4 concentric-unit model produces an accurate simulation of the illusory circle oscillations observed on viewing the Marroquin pattern (Wilson et al., 2000a). In particular, the model oscillations were well described by a Gamma distribution with $\lambda = 1.6$ and $n = 1.7$, which are very close to the values obtained from human subjects. A simulation of the illusory circle dynamics generated by this V4 model network may be viewed on the World Wide Web at http://neurosci.nature.com/web_specials/.

How might this competitive V4 concentric unit network explain aspects of selective attention? First, the network architecture incorporates the inhibitory competition that is believed to form the neural basis of attention (Desimone and Duncan, 1995). Second, a novel form of attentional biasing suggests itself here. The ion currents causing spike frequency adaptation are known to be under neuromodulatory control (McCormick and Williamson, 1989). In particular, histamine, acetylcholine, norepinephrine, and serotonin all reduce the magnitude of spike frequency adaptation in excitatory cortical neurons. A spatially localized reduction in spike frequency adaptation could readily be introduced into the V4 network via top-down neuromodulatory transmitter release. This would produce a bias in the network, conferring a competitive advantage on those neurons exhibiting less spike frequency adaptation as a result of neuromodulation and therefore a higher asymptotic firing rate. Such a reduction of spike frequency adaptation would have a small effect on the initial neural transient but a much greater effect on the subsequent asymptotic firing level, and this is precisely the time course of adaptational effects reported for single neurons in primate V4 (McAdams and Maunsell, 1999). Thus, we hypothesize that neuromodulation may play the key role in implementing top-down attentional network biasing (Wilson et al., 2000a).

Although we have focused here on concentric units, there are certainly many categories of V4 configural units. Configural biasing can no doubt be applied to one or another class of these in order to implement feature-based attention. However, concentric V4 units have properties that recommend them as a substrate for spatial and object based attention. As shown in Fig. 4.3 earlier, the activity of a model V4 concentric unit locates and measures the diameter of an ellipsoidal region in the retinal image. In natural images such areas frequently delimit biologically relevant objects such as heads, fruit, or animal torsos. Selection of ellipsoidal regions by a V4 network may thus serve to attentively select candidate objects for further scrutiny. In this regard, recall that the model V4 network effectively extracted head location and size from the stimulus in Fig. 4.4, and studies using such face-house combinations have provided evidence for objects as the basis for selective attention (O'Craven, Downing, and Kanwisher, 1999). Finally, feedback from activated V4 concentric units to retinotopically corresponding areas in V1 could provide the substrate for the figure-ground contextual modulation discovered in V1 by Zipser, Lamme and Schiller (1996). Analogous comments no doubtlessly apply to the activity of other categories of V4 configural units.

4.10   Summary and Overview

Intermediate level form vision clearly involves multiple components. We have focused on just one of them - configural grouping of orientation information. Based on a consideration of receptive field sizes we argue that this process is likely to occur through a series of sequential operations that involve at least three cortical regions in which radial frequency dimensions show the necessary systematic increase. Our psychophysical results provide strong evidence for the existence of at least two basic configural organizations in intermediate vision: circular and radial. Networks of the former provide a means of encoding a wide range of ellipsoidal shapes, whereas individual units with radial organization encode cross or X-shaped structures. (We have not yet examined the network capabilities of such radial units.) There is also evidence physiologically for hyperbolic organization, and this may not yet be an exhaustive list. On the other hand, we doubt that the number of basic configural organizations at this level is enormous. Because the possible subtypes must be duplicated – probably at multiple spatial scales – across the visual field, and cortical areas beyond V2 have much smaller overall areas than V1 and V2, a compact code is essential.

Our psychophysical modelling and fMRI findings point to the existence of an intermediate form-processing region in the human brain which plays a role analogous to primate V4. From the perspective of our research, V4 configural units play a major role in the description and measurement of stimulus shapes. Such measurement and description goes beyond contour enhancement by collinear facilitation in V1. Thus, collinear facilitation will enhance the saliency of smoothly varying contours encoded by V1 simple cells, whereas V4 units identify and measure such biologically relevant contour configurations as circles, ellipses, crosses, etc. Beyond V4 we envisage flexible pooling of V4 configural unit responses plus pooling across spatial scales as giving rise to size-invariant representations of biologically relevant objects including faces, fruits, tools, and alphanumeric characters. Such configural unit response pooling would explain the increasing receptive field sizes in TEO and TE in the same spirit as has been developed for V1, V2, and V4 in Fig. 4.2. Unravelling the details of these further transformations leading to object recognition will require cooperation among primate physiologists, human psychophysicists, brain imaging, and modelers.

Acknowledgments

This work was supported in part by grants from NSERC to FW (OGP0007551) and NIH to HRW (EY02158).

References

Allison, T., Puce, A., Spencer, D. D. and McCarthy, G. (1999). Electrophysiological studies of human face perception. I: Potentials generated in occipitotemporal cortex by face and non-face stimuli., *Cerebral Cortex*, 9:415-430.

Blake, R. (1989). A neural theory of binocular rivalry, *Psych. Rev.*, 96:145-167.

Boussaoud, D., Desimone, R. and Ungerleider, L. G. (1991). Visual topography of area TEO in the macaque, *J. Comp. Neurol.*, 306:554-575.

Burbeck, C. A. and Pizer, S. M. (1995). Object representation by cores: Identifying and representing primitive spatial regions, *Vis. Res.*, 35:1917-1930.

Connor, C. E., Preddie, D. C., Gallant, J. L. and Van Essen, D. C. (1997). Spatial attention effects in macaque area V4, *J. Neurosci.*, 17:3201-3214.

Connors, B. W. and Gutnick, M. J. (1990). Intrinsic firing patterns of diverse neocortical neurons, *Trends Neurosci.*, 13:99-104.

De Weerd, P., Peralta, M. R., Desimone, R., and Ungerleider L. G. (1999). Loss of attentional stimulus selection after extrastriate cortical lesions in macaques, *Nature Neurosci.*, 2:753-758.

Desimone, R. and Duncan, J. (1995). Neural mechanisms of selective visual attention, *Ann. Rev. Neurosci.*, 18:193-222.

Desimone, R. and Schein, S. J. (1987). Visual properties of neurons in area V4 of the macaque: sensitivity to stimulus form, *J. Neurophysiol.*, 57:835-868.

DeValois, R. L., Albrecht, D. G., and Thorell, L. G. (1982). Spatial frequency selectivity of cells in macaque visual cortex, *Vis. Res.*, 22:545-559.

Dow, B. M., Snyder, A. Z., Vautin, R. G. and Bauer, R. (1981). Magnification factor and receptive field size in foveal striate cortex of the monkey, *Exp. Brain Res.*, 44:213-228.

Felleman, D. I. and Van Essen, D. C. (1991). Distributed hierarchical processing in the primate cerebral cortex, *Cerebral Cortex*, 1:1-47.

Felleman, D. J., Xiao, Y. and McClendon, E. (1997). Modular organization of occipito-temporal pathways: cortical connections between visual area 4 and visual area 2 and posterior inferotemporal ventral area in macaque monkeys, *J. Neurosci.*, 17:3185-3200.

Fox, R., and Herrmann, J. (1967). Stochastic properties of binocular rivalry alternations, *Percept. and Psychophys.*, 2: 432-436.

Gallant, J. L., Braun, J. and Van Essen, D. C. (1993). Selectivity for polar, hyperbolic, and Cartesian gratings in macaque visual cortex, *Science*, 259:100-103.

Gallant, J. L., Connor, C. E., Rakshit, S., Lewis, J. W. and Van Essen, D. C. (1996). Neural responses to polar, hyperbolic, and Cartesian gratings in area V4 of the macaque monkey, *J. Neurophysiol.*, 76:2718-2739.

Gattass, R., Gross, C. G. and Sandell, J. H. (1981). Visual topography of V2 in the macaque, *J. Comp. Neurol.*, 21:519-539.

Gattass, R., Sousa, A. P. B. and Gross, C. G. (1988). Visuotopic organization and extent of V3 and V4 of the macaque, *J. Neurosci.*, 8:1831-1845.

Glass, L. (1969). Moiré effect from random dots, *Nature*, 223:578-580.

Glass, L. and Pérez, R. (1973). Perception of random dot interference patterns, *Nature*, 246:360-362.

Gross, C. G. (1992). Representation of visual stimuli in inferior temporal cortex, *Phil. Trans. Roy. Soc. Lond.*, 335:3-10.

Gross, C. G., Bender, D. B. and Rocha-Miranda, C. E. (1969). Visual receptive fields of neurons in inferotemporal cortex of the monkey, *Science*, 166:1303-1306.

Heeger, D. J. (1992). Normalization of cell responses in cat striate cortex, *Vis. Neurosci.*, 9: 181-197.

Hess, R. F., Wang, Y.-Z. and Dakin, S. C. (1999). Are judgements of circularity local or global?, *Vis. Res.*, 39:4354-4360.

Hubel, D. H. and Wiesel, T. N. (1968). Receptive fields and functional architecture of monkey striate cortex, *J. Physiol.*, 195:215-243.

James, T. W., Wilkinson, F., Wilson, H. R., Gati, J. S., Menon, R. S. and Goodale, M. A. (1999). fMRI activation of fusiform face and object areas by concentric and radial gratings, *Invest. Ophthal. Vis. Sc.*, 40:S819.

Kimia, B. B., Tannenbaum, A. R. and Zucker, S. W. (1995). Shapes, shocks, and deformations I: the components of two-dimentsional shape and the reaction-diffusion space, *Int. J. Comp. Vis.*, 15:189-224.

Livingston, M. S. and Hubel, D. H. (1984). Anatomy and physiology of a colour system in the primate visual cortex, *J. Neurosci.*, 4:309-356.

Lorenzon, N. M. and Foehring, R. C. (1992). Relationship between repetitive firing and afterhyperpolarizations in human neocortical neurons, *J. Neurophysiol.*, 67:350-363.

Luck, S. J., Chelazzi, L., Hillyard, S. A. and Desimone, R. (1997). Neural mechanisms of spatial selective attention in areas V1, V2 and V4 of macaque visual cortex, *J. Neurophysiol.*, 77:24-42.

Marroquin, J. L. (1976). Human visual perception of structure, Unpublished Master's Thesis, Massachusetts Institute of Technology.

Maunsell, J. H. and Van Essen, D. C. (1983). The connections of the middle temporal visual area (MT) and their relationship to a cortical heirarchy in the macaque monkey, *J. Neurosci.*, 3:2563-2586.

McAdams, C. J. and Maunsell, J. H. R. (1999). Effects of attention on orientation tuning functions of single neurons in macaque cortical area V4, . *Neurosci.*, 19:431-441.

McCormick, D. A. and Williamson, A. (1989). Convergence and divergence of neurotransmitter action in human cerebral cortex, *Proc. Nat. Acad. Sci.*, 86:8098-8102.

Meadows, J. C. (1974). The anatomical basis of prosopagnosia, *J. Neurol., Neurosurgery and Psychiatry*, 37:489-501.

Moran, J. and Desimone, R. (1985). Selective attention gates visual processing in the extrastriate cortex, *Science*, 229:782-784.

Moscovitch, M., Winocur, G. and Behrmann, M. (1997). What is special about face recognition? Nineteen experiments on a person with visual object agnosia and dyslexia but normal face recognition, *J. Cog. Neurosci.*, 9:555-604.

Motter, B. C. (1993). Focal attention produces spatially selective processing in visual cortical areas V1, V2 and V4 in the presence of competing stimuli, *J. Neurophysiol.*, 70.

Motter, B. C. (1994). Neural correlates of attentive selection for color or luminance in extrastriate area V4, *J. Neurosci.*, 14:2178-2189.

O'Craven, K. M., Downing, P. E. and Kanwisher, N. (1999). fMRI evidence for objects as the units of attentional selection, *Nature*, 401:584-587.

Reynolds, J. H., Chelazzi, L. and Desimone, R. (1999). Competitive mechanisms subserve attention in macaque areas V2 and V4, *J. Neurosci.*, 19:1736-1753.

Stevens, K. A. and Brookes, A. (1987). Detecting structure by symbolic constructions on tokens, *Comp. Vis. Graphics and Image Proc.*, 37: 238-260.

Tsotsos, J. K. (1990). Analyzing vision at the complexity level, *Behav. and Brain Sci.*, 13:423-469.

Ungerleider, L. G. and Mishkin, M. (1982). Two cortical visual systems, in D. J. Ingle, M. J. Goodale and R. J. W. Mansfield (Eds.), *Analysis of Visual Behavior*, 549-586, Cambridge, MA: MIT Press.

Van Essen, D. C. and Drury, H. A. (1997). Surface and functional analyses of human cerebral cortex using a surface-based atlas, *J. Neurosci.*, 17:7079-7102.

Van Essen, D. C., Newsome, W. T., and Maunsell, J. H. R. (1984). The visual field representation in the striate cortex of the macaque monkey: asymmetries, anisotropies, and individual variability, *Vis. Res.*, 24:429-448.

von der Heydt, R., and Peterhans, E. (1989). Mechanisms of contour perception in monkey visual cortex. 1. Lines of pattern discontinuity, *J. Neurosci.*, 9:1731-1748.

von der Heydt, R., Peterhans, E., and Baumgartner, G. (1984). Illusory contours and cortical neuron responses, *Science*, 224:1260-1262.

Wilkinson, F., Wilson, H. R. and Habak, C. (1998). Detection and recognition of radial frequency patterns, *Vis. Res.*, 38:3555-3568.

Wilson, H. R. (1999a). Non-Fourier cortical processes in texture, form, and motion perception, in P. Ulinski, and E. G. Jones (Eds.), *Cerebral Cortex, Vol. 13: Models of Cortical Circuitry*, 445-477, New York, NY: Plenum.

Wilson, H. R. (1999b). *Spikes, Decisions, and Actions: Dynamical Foundations of Neuroscience*, Oxford: Oxford University Press.

Wilson, H. R., Krupa, B., and Wilkinson, F. (2000a). Dynamics of perceptual oscillations in form vision, *Nature Neurosci.*, 3:170-176.

Wilson, H. R. and Wilkinson, F. (1998). Detection of global structure in Glass patterns: implications for form vision, *Vis. Res.*, 38:2933-2947.

Wilson, H. R., Wilkinson, F. and Asaad, W. (1997). Concentric orientation summation in human form vision, *Vis. Res.*, 37:2325-2330.

Wilson, H. R., Wilkinson, F., Lin, L.-M. and Castillo, M. (2000b). Perception of head orientation, *Vis. Res.*, 40:459-472.

Zeki, S. M. (1971). Cortical projections from two prestriate areas in the monkey, *Brain Res.*, 34:19-35.

Zipser, K., Lamme, V. A. F. and Schiller, P. H. (1996). Contextual modulation in primary visual cortex, *J. Neurosci.*, 16:7376-7389.

5

Visual Attention: The Active Vision Perspective

John M. Findlay
Iain D. Gilchrist

This chapter discusses the relationship between covert visual attention and overt movements of the eyes. One of the most frequently emphasised facts concerning visual attention is the ability to attend covertly to a location in the visual field without directing the eyes to that location. Based on this, the "mental spotlight" metaphor has been widely prevalent and has given rise to much experimental work. An important, but rather neglected, question concerns what role such a mental-spotlight process might play in vision. The compelling nature of the spotlight metaphor seems likely to arise because of its similarity with overt movements of the eyes. This leads to the suggestion that covert attention is involved in some form of mental scanning. We will argue that this account is not satisfactory. An alternative account, termed active vision, develops from the fact that overt eye scanning occurs several times each second in many tasks. Analysis of information intake during text reading, scene perception and visual search suggests that no additional covert scanning occurs in these cases. Evidence however suggests that attentional processes may operate to assist in pre-processing information in the visual periphery at the location to which the eyes are about to be directed. Covert attention to peripheral locations thus acts to supplement, not substitute for, actual movements of the eyes.

5.1 Introduction

There was an extraordinary proliferation of work on visual attention in the late 1990's, summarised in several recent books (Styles, 1997; Pashler, 1998; Wright, 1998) and numerous journal articles (reviews include Egeth and Yantis, 1997; Cave and Bichot, 1999). A frequent starting point in discussions is the simple demonstration that attention to visual locations can operate covertly. An observer can, while fixating one location, attend to a different location without moving the eyes to look at it. This result, known to Helmholtz, was brought into prominence by the work of Michael Posner (Posner, Nissen, and Ogden, 1978; Posner, 1980) whose experimental paradigms have proved highly productive. Here we consider how covert attention is used in common visual tasks. We argue that covert attention does not operate as an independent scanning mechanism but instead supplements the process of overt eye scanning that forms the normal way that visual attention is redirected.

Attention to a location demonstrably affects the speed and accuracy of visual processing at that location. Visual material at the attended location is selected for processing and receives enhanced processing. Attention generally leads to benefits but an ingenious demonstration by Yeshurun and Carrasco (1998) describes a case where attention actually renders visual discriminations less effective. Two separate preconditions are recognised. Attention may be directed voluntarily in response to instructions or cues. However, attention may also be captured by environmental events. In particular, the occurrence of visual change in the periphery has the effect of summoning attention to its location. The consequences, in terms of enhanced processing of subsequent material, are similar in the two cases, but differences in time course and other properties are found (Müller and Rabbitt, 1989; Nakyama and Mackeben, 1989). These two forms of attention are distinguished in various ways (endogenous, exogenous; sustained, transient etc). At the heart of this chapter is the question: How are the attentional effects best described? Posner and followers have emphasised a spotlight metaphor whereby attention is envisaged as a mental process akin to a spotlight beam which "illuminates" a restricted region of some internal mental representation of the visual scene. Much work on visual attention has been concerned with exploring the validity of the spotlight account and offering possible alternatives. The suggested alternatives (zoom lens, hemifield inhibition, objects rather than locations), in general, retain the idea that attention operates covertly by facilitating some part of the mental representation at the expense of others.

The implicit assumption of the spotlight model is the existence of a full and rich mental representation upon which this attentional spotlight could act and across which it could move. Until recently, this assumption had been unquestionable. It appeared self-evident that the spotlight had something to illuminate. Vision was analysed as a process that started at the retinal image. The work of David Marr (1982) epitomised this approach and offered a computational account elaborating processes that might transform a raw retinal image into a three-dimensional scene representation. The 1990s saw the gradual realisation that this account of vision may be seriously misleading. As pointed out by O'Regan (1992, see also Nakayama, 1992), the traditional image-processing approach did not appear to be nearing a solution of "the real mysteries of vision". A polemic entitled "A critique of pure vision" (Churchland, Ramachandran, and Sejnowski, 1994) argued that the "picture-in-the-head" metaphor for vision was still too pervasive among vision scientists and had outlived its usefulness. The incompleteness of the mental visual representation has now become widely accepted, and more recent work on "change blindness" (Grimes, 1996; Rensink, O'Regan and Clark, 1997; Rensink, 2000) has provided a convincing immediate demonstration.

These developments necessitate a reappraisal of the role of attention in the visual process. In this chapter we will examine vision from a somewhat different perspective – that of active vision – and discuss how attention might be incorporated into this framework.

5.2   Active Vision

Active Vision is a term that appears to have been coined by computer scientists working on visual problems. For a time, work on computer vision was dominated by the image-processing approach advocated by Marr (1982). Attempts were made to derive an elaborated representation at each location of a visual scene. These attempts were largely unsuccessful and led to an appreciation of the vast computational resources that would be entailed in order to derive a full three-dimensional representative description across the whole of an extended image. A promising alternative emerged from work that downplayed the perceptual aspect of vision and concentrated on the means whereby a vision system might lead to useful visually controlled behaviour. A key to progress appeared to be a visual sensor, of which a region, effectively a fovea, could be directed to different locations within the visual field. Such a system allows vision to deploy *deictic*, or pointing, properties that are essential for interfacing with the cognitive system (Ballard, Hayhoe, Pook and Rao, 1998). Aloimonos, Bandopadhay and Weiss (1987) termed such a system an *active vision system*. Ballard (1991) elaborated very similar ideas but preferred the term "*animate vision*[1]". Of course, this work developed in full awareness that the vision of humans, and that of other active animals, utilises the mobility of the eyes (Land, 1995). In many situations, rapid saccadic, or jump-like, movements reposition the gaze several times each second. No useful vision occurs during these movements and information intake is restricted to the successive fixations of the eye, pauses of a fraction of a second between the scanning movements. These movements are characteristic of most daily visual activity. Land, Mennie, and Rusted (1999), in an elegant study, comment that "the eye-movement [record] has the frenetic appearance of a movie that has been greatly speeded up."

It is something of a mystery that textbooks of vision frequently ignore the mobility of the eye. Equally disturbing is the tendency to ignore the fundamental nonhomogeneity of the visual system that renders these movements essential. The highest visual acuity is only achieved in the region of the central fovea. The fovea is technically defined as an anatomical depression in the retina occupying about 2 degrees. In human vision, this is also, at least approximately, the area from which rods are absent and vision is mediated entirely by cones. However, while the fovea can be defined in anatomical terms, there is no corresponding functional boundary at its edge. Visual acuity declines steadily from the very centre of the fovea out to the periphery (Wertheim, 1894), as do many other visual functions (Rovamo,

---

[1]Ballard wished to avoid possible confusion with active sensing, a term preempted in the computer-vision world. Active vision might also offer confusion with the quite different approach of J. J. Gibson (1966, etc.). Nevertheless, we believe it is the term that most appropriately captures the approach.

Virsu and Näsänen, 1978).[2] The nature of this steady decline gives it a remarkable self-scaling property. As remarked by Anstis (1974), the acuity vs eccentricity tradeoff function ensures that the visibility of visual material is largely unaffected by its retinal size. Retinal size is affected by viewing distance, and the self-scaling property is surely of significance in allowing us to adapt our visual behaviour to material at a wide variety of viewing distances.

The active-vision approach emphasises the sequential nature of information intake in vision. The important questions become: What information is taken in at each fixation? How is this information integrated with that from preceding and subsequent fixations? How are the scanning movements planned and orchestrated? In one specialised area – the study of information intake during reading – much progress has been made towards answering these questions (Rayner, 1998). The reading process has some unique simplifying constraints. The eyes move across the text, in general, in a prespecified direction, and acuity limitations mean that relatively little use is made of parafoveal or peripheral vision. Other situations are less constrained and less understood, although some progress is being made in the areas of scene perception and search (Henderson and Hollingworth, 1999; Liversedge and Findlay, 2000).

How might covert attentional processes interact with overt scanning movements of the eye? Several possibilities exist. First, covert attention might be entirely absent when overt eye scanning occurs. Covert attention might be just a curious artefact of the peculiar instructions given in the psychology laboratory. Second, at the opposite extreme, overt eye movements might be incidental. Covert attention shifts might be the main way in which attention was redirected. In some accounts, it is suggested that several locations might be scanned by covert attention within each eye fixation. For this sequential covert attentional scanning to occur within a single fixation it would, of course, be necessary for covert attention to be redeployed much more rapidly than the speed of overt eye scanning.

We will argue that neither of these extreme positions is tenable, and particularly argue against the existence of "sequential intrafixational scanning". A more promising approach builds on the long recognised link between covert attention and eye movement programming. It was demonstrated some years ago (Shepherd, Findlay and Hockey, 1986; see also Hoffman and Subramaniam, 1995) that during the preparatory period of a voluntary eye movement, responses to an attentional probe were faster at the destination location of the eye movement, even when the manipulation of probe probability had rendered an alternative location more probable. Such a finding is consistent with the *pre-motor theory of attention* (Rizzolatti, Riggio, DaScola, and Umiltá, 1987; Rizzolatti, Riggio and Sheliga, 1994). This theory proposes that covert attention to a visual location involves using the mechanisms for saccade programming, but with the actual motor response withheld.

---

[2]Despite this fact, we will continue to use the convenient terms fovea, parafovea, and periphery as introduced by Rayner (1984). "Foveal vision" refers to the central 2 deg of vision and "parafoveal vision" refers to the central 10 deg with "periphery" referring to the remainder. In relation to the total visual field, the fovea occupies about 0.04% of the total area and the parafovea about 1%.

We believe the pre-motor theory is essentially correct but that, rather than putting the emphasis on pre-motor theory as an account of covert attention, it should be recognised that overt attention has primacy.

Further evidence indicating this attention link comes from experiments showing that visual discriminations at the destination location of an eye movement are improved in comparison with other locations (Kowler, Anderson, Dosher and Blaser, 1995; Deubel and Schneider, 1996). These two findings demonstrated that the improved discrimination occurred before the eye movement itself commenced. Many studies have demonstrated *pre-view advantage* when saccadic eye movements are made. The visibility of material in parafoveal or peripheral vision prior to an eye movement facilitates subsequent recognition. This suggests that the improved pre-saccadic visual discrimination at the saccade destination is normally incorporated into an integrated routine for dealing with information trans-saccadically, thereby speeding up the recognition process.

This facilitation of discriminations at the destination location of a saccade is often described by saying that covert attentional processes precede the saccadic movement. It is further frequently assumed that, during the preparatory period of the saccade, covert attention moves from the fixated location to the saccade destination location in a spotlight-like manner. However, this extension assigns concrete properties to attention, such as a unique and restricted locus. As Kowler et al. (1995) point out, this assumption may be unwarranted. One tradition in attentional work has argued that spatial shifts of attention between two locations are characterised by three mental operations; disengagement of attention at the first location, movement of attention, and re-engagement of attention at the second location (Posner, Walker, Friedrich, and Rafal, 1984). Although superficially plausible, this position may be challenged. Thus one piece of supporting evidence, the "gap effect" in saccadic eye-movement programming, has proved better interpreted in an alternative way (Kingstone and Klein, 1993; Findlay and Walker, 1999).

We will now consider the role of attention in turn in the areas of reading, scene and object perception, and visual search.

5.3   Reading

Reading is a ubiquitous and very familiar activity. The study of visual information intake in reading has a long history and provides a benchmark situation for studies of active vision. An excellent review by Rayner (1998) gives a very thorough general account. We will select certain details here, which emphasise possible involvement of visual attention in the reading process.

A reader typically advances along a line of text making a sequence of fixations and saccadic movements. The average length moved is around 7–8 characters, showing a rough correspondence to the average length of a word. Short words are frequently not fixated and long words will frequently receive more than one fixation. Occasionally a regressive movement in the reverse direction breaks the forward sequence of the scanning.

An extremely fruitful innovation for investigating the reading process was the *gaze-contingent methodology* developed by McConkie and Rayner (1975). This combined advances in computer tecl:nology with advances in eye gaze recording to present observers with visual displays where the visual information could be changed in a manner dependent on where the observer's gaze was directed. Various different detail manipulations of this generic technique have been developed, among which the *moving window* manipulation is perhaps the most widely used. In the moving window technique, as applied to reading, the subject sees normal text within a window of predefined size but outside this window the text is masked in some way, for example each text character may be replaced by an "x." Each time the eye moves, the display is changed so that the unmodified material is always presented precisely where the gaze is directed.

This technique has been used to measure the *perceptual span* in reading. If the manipulation does not affect the speed of reading, it is reasonable to assume that the material outside the window is playing no part in the normal process. Conversely, when the window size is reduced so that reading speed is affected adversely, then it can be deduced that material from regions outside the boundary must normally be processed. This allows measurement of the perceptual span, the region from which visual information is extracted. Summarising many results, the perceptual span is found to extend no more than 15 characters to the right of fixation and only 3–4 characters to the left. Information about the visual characteristics of letters is only available within a smaller window, extending 8–12 characters to the right; the contribution to the span of material more peripheral than this comes exclusively from word boundaries.

The asymmetry of the span is an immediate indication that attentional effects are involved. Visual acuity declines systematically away from the fovea with no left-right differences. The asymmetry has been found to reverse when readers read text in languages such as Hebrew, where the direction of reading goes from right to left. This finding alone allows rejection of one of the suggestions made earlier, that attentional processes are not operative when free eye scanning is allowed.

A study by Blanchard, McConkie, Zola, and Wolverton (1984) addressed the possibility that, during each fixation, a covert attentional scanning process operates within the material in the perceptual span. For example, suggestions had been made that reading operates as a sequential letter-by-letter serial encoding process (Gough, 1972). Blanchard et al. programmed a gaze-contingent display where the display changed three times during each fixation. An example is shown in Fig. 5.1.

This sequence of display changes occurred on every fixation. The critical change occurred to one carefully chosen five letter word of the text, where one letter was changed from its first exposure (time T_1) to its second (time T_2). Thus, during the early part of the fixation (T_1), the first word ("tombs") is present, whereas during the second part (T_3), the second word ("bombs") has replaced it. Equivalent changes were made where letters at different positions in the word sequence (e.g., changing

	*
T$_1$	the underground chambers were meant to hide hidden tombs, but then the construction
T$_2$	xxxxxxxxxxxxxxxxxxxxxxxxxxxxxxxxxxxxxxxxxxxxxxxxxxxxxxxxxxxxxxxxxxxxxxxxxxxx
T$_3$	the underground chambers were meant to hide hidden bombs, but then the construction

FIGURE 5.1. Text changes in the experiment of Blanchard et al. (1984). The experimental subject read a line of text. During every fixation, the text was changed at the critical point marked by the asterisk. Up to time T$_1$, the word "tombs" was displayed. After a further time T$_2$, the word "bombs" replaced it. The brief masking stimulus presented (for 30 msec) at time T$_2$ prevented the attention-grabbing effect that would otherwise occur when the letter was changed.

the word "tombs" to "tomes") were made. A critical prediction of the left-to-right serialisation hypothesis is an effect of the relative locations of fixation and critical letter on the probabilities of reporting the first or the second word. When the changed letter occurs to the left of the fixation location, then the first word should be more likely to be reported, whereas when the critical letter lies to the right, an increase in the probability of reporting the second letter is expected. The results did not provide any evidence to support the attentional-scanning prediction. Subjects would sometimes report the first word and sometimes the second[3]. The likelihood of reporting each word depended on the timing of the changes (times T$_1$ and T$_3$) but was not affected by the relative fixation position.

The results of the Blanchard et al. (1984) study suggest that a sequential left-to-right intake of letters does not occur. Such a possibility is an extreme form of the sequential intrafixational scanning position. It could only occur if covert attention could scan the letters at a very rapid rate. Rates of attentional scanning will be considered further in a subsequent section, and it will be argued that there is no good evidence that scanning of attentional focus can occur at a rate that is considerably in excess of the rate of overt scanning. We consider next a proposal that has widespread, although not universal, support. This is the idea that within a fixation, the attentional focus is initially on the fixated word and moves at some point during the fixation to the following word in the text.

The length of the perceptual span is such that, frequently, two words are present within the span. If the moving window is reduced so that only one word is available, reading is possible but with some cost. Inhoff and Rayner (1986) found that gaze durations on a word were about 30 msec shorter when the word had been previously visible in the periphery of vision than when the word had not been previewed. The usual interpretation of this finding is that during a fixation, information is taken in, both from the word fixated, and also from the following word (on the right of fixation for English readers). This information can be used to speed up some process, possibly lexical access, at the time when the following word is, in its turn,

---

[3]On a small proportion of occasions (35%), both words could be reported. These trials were excluded from subsequent analyses.

fixated. It may also be used in skipping the next word. The term *peripheral preview advantage* succinctly describes the result. Peripheral preview in reading-like tasks has been considerably investigated (Rayner, 1998). A principal concern has been what type of information is extracted and integrated across the saccade that brings the word into foveal vision. Rayner (1998) reviews the evidence and suggests that abstract letter codes and phonological codes are used for this purpose.

The peripheral-preview effect is well established and shows the operation of attentional processes in reading. However there are two alternative positions concerning how these attentional operations are effected. One position argues that covert attention is located at the fovea during the initial part of the fixation and, at some point during the fixation, shifts to the destination of the subsequent saccade. The second position argues that during the fixation, material from the fovea and the saccade destination location is processed in parallel. The first position has been advocated in a series of models of increasing sophistication produced by the Amherst group (Morrison, 1984; Henderson and Ferreira, 1990; Reichle, Pollatsek, Fisher, and Rayner, 1998). The second position is linked to the work of McClelland and Mozer (1986).

The idea that attention shifts at a specific point in time prior to the actual eye movement appears intuitively plausible but it is difficult to support with direct evidence. It has been shown that foveal information can be extracted if only visible for the first 50 msec of the fixation (Slowiaczek and Rayner, 1987). However the study of Blanchard et al. (1984) discussed earlier shows that information presented at a later point during the fixation is also processed. Other tests involve further assumptions. The models that involve a discrete attention shift identify a process that triggers this shift of attention. This is generally related to the lexical processing of the foveated word. A testable prediction is thus generated. Because the processing of the foveal word determines when attention shifts, then the content of the parafoveal word cannot affect the timing of this attention shift. If the further assumption is made that the attention shift is followed after a fixed time by the actual eye movement, neither will it affect the fixation time. Henderson and Ferreira (1993) tested this prediction and reported no effect of the processing difficulty of the parafoveal word on fixation times. However, Kennedy (1998; see also Inhoff, in press) presents a contrary finding.[4]

In summary, work on reading has very clearly demonstrated attentional effects through the asymmetry of the perceptual span. It is clear that reading is characterised by active sampling of visual information with the eyes and that a small but significant contribution to the process comes from a preview of the material to which the eyes are to be directed next. There is still debate about whether this preview occurs because the attentional focus makes a specific movement during the

---

[4]It may be noted that there is a further lively debate about how the extent of parafoveal preview depends on the difficulty of the fixated word (Henderson and Ferreira, 1990; Kennison and Clifton, 1995; Schroyens, Vitu, Brysbaert and d'Ydewalle, 1999). Although this has relevance for theories involving attentional movement, it brings in the "limited-capacity" view of attention, and a full discussion would go beyond the scope of this chapter.

fixation, or whether the foveal and the attended parafoveal material are processed in parallel. Importantly for the current chapter, there is no support for a more e-laborated sequence of covert attentional movements (for example a letter-by-letter scan) within a fixation.

5.4   Scenes and Objects

Although much less intensively investigated, some of the principles emerging from studies of active vision in reading are also at work during the more general situation of viewing visual material. An important difference is that scenes allow "gist" information to be extracted very rapidly. The nature of this gist information has been a topic of considerable recent interest, in view of the demonstrations (change-blindness) that much less specific information is available than once thought. One suggestion is that gist information relates to statistical properties of colour and contour distribution rather than more fully analysed information (Henderson and Hollingworth, 1999).

Attempts have been made to describe the useful field of view for picture percep-tion using a gaze-contingent window technique in a manner similar to that used in studies of reading. Saida and Ikeda (1979) created a viewing situation with an electronic masking technique in which subjects saw only a central square region of an everyday picture. Viewing time and recognition scores were impaired unless the window was large enough for about half the display to be visible. This fraction applied both to 14 deg by 18 deg displays and to 10 deg by 10 deg ones. A follow up study by Shiori and Ikeda (1989) looked at the situation where the picture was seen normally inside the window but in reduced detail outside. Even very low-resolution detail (very well below the acuity limit at the peripheral location) aided performance considerably. Shiori and Ikeda used a pixel-shifting technique to re-duce resolution. Van Diepen, Wampers, and d'Ydewalle (1998) have developed a technique whereby a moving window can be combined with Fourier filtering. Somewhat surprisingly, performance was better when the degraded peripheral ma-terial was low-pass filtered than when it was high-pass filtered.

When scenes are viewed containing a number of different objects, fixations occur on these objects and are important. Nelson and Loftus (1980) allowed subjects a limited amount of time to scan such a scene, and they then performed a recognition test, viewing a version of the scene in which one object had been changed. Their task was to detect the changed object. The detection rate for small objects (1 deg) was above 80% for objects that had been directly fixated and higher if they received two fixations. However, it fell to 70% for objects where the closest original fixation had been at a distance from the object in the range 0.5 – 2 deg and to a chance level (or in one case slightly above chance) for objects viewed more peripherally.

As with words, parafoveal preview aids object identification. A series of studies at Amherst University has used the boundary technique as shown in Fig. 5.2 to explore the preview benefit of presaccadic material in peripheral vision. Pollatsek,

Fixation 1   +

Fixation 2   +

FIGURE 5.2. The experiment of Pollatsek et al. (1984). On fixation 1, visual material was presented in peripheral vision, which the subject was required to fixate and name. During the saccade prior to the second fixation, the material could be changed in various ways or left unchanged.

Rayner and Collins (1984) measured the speed of object naming when various forms of preview occurred, using as a baseline a square containing a cross. They found that full peripheral preview produced a substantial benefit (100–130 msec). Benefits, albeit of a reduced magnitude (90 msec), also occurred if the previewed picture came from the same category as the target picture. A small benefit (10 msec) occurred when the previewed picture had the same overall visual shape as the picture (e.g., carrot > baseball bat). This preview information only occurs for previewed material at the saccade destination: material elsewhere in the visual field does not generate any priming (Henderson, Pollatsek and Rayner, 1989).

The emerging view from this discussion is that scene perception, apart from the initial gist, is built up from the serial eye scan. A result apparently in contradiction to this position was reported by Loftus and Mackworth (1978). They asked people to look at scenes, some of which contained an object that was highly unlikely in the context (e.g., an octopus in a farmyard). Control scenes had a visually similar but contextually probable object (a tractor in a farmyard), and subjects' eye scans were recorded. Loftus and Mackworth reported that the anomalous object was detected rapidly and often when the eyes were located in a remote part of the scene immediately prior to the movement to the anomaly. This implied that detailed analysis of scene content in peripheral vision was occurring and influencing subsequent behaviour. However, as reviewed by Henderson and Hollingworth (1998, 1999), several attempts to replicate Loftus and Mackworth's findings have now been attempted and all have failed to replicate the crucial findings.

Work on object and scene perception shows that peripheral vision is much more important in this area than in reading. However, the identification of objects within scenes appears to proceed in a way quite similar to the identification of words during reading. Peripheral preview processes are likely to be operative, but none of the current work on scene perception makes use of the concept of spotlight-based shifts of covert attention.

5.5   Search

In visual search, the task is to locate a target on the basis of some visual properties. Everyday examples of visual-search tasks abound (finding a cup in the kitchen, finding a book on a bookshelf). The target book may be specified precisely (Harris and Jenkin's *Vision and Action*) or loosely (a red object). It is also easy to set up controlled laboratory tasks to investigate visual search. Typically a display is presented containing a target together with a number of additional items, termed nontargets or distractors.

A dominant tradition in visual search was initiated with a seminal paper by Treisman and Gelade (1980). They argued that some primary visual properties allow a search in parallel across large displays. In such cases the target appears to 'pop out' of the display. For example there is no problem in searching for a red item amongst distractor items coloured green, blue or yellow. In other cases, the paradigm example being a 'feature conjunction' search for a target that is both green and a cross when distractors include red crosses and green circles, the task is much more difficult, suggesting the use of a different search strategy. Treisman and Gelade argued that in the pop-out tasks preattentional mechanisms permit rapid target detection, in contrast to the conjunction task, which was held to require a serial deployment of attention over each item in turn.

Treisman and Gelade introduced an experimental paradigm that differentiated the different types of searches. They measured the time taken for an observer to make a speeded two choice decision concerning the presence or absence of a target in a visual display. Half the displays contained a target and in the remaining the target was absent. The critical independent variable was the number of display items. The *search function* shows how the response time depends on this variable. Figure 5.3 shows the classic search functions from Experiment 1 of Treisman and Gelade's paper.

The traditional interpretation of the search function is that the display-size-dependent increases shown in the search functions for conjunction searches come about through an item by item serial scan of covert attention through the display. If the display does not contain a target, it is assumed that every item in the display is scanned before a target-absent response is given. If the search is self-terminating in displays that do contain a target, then on average half the display items must be scanned before the target is found. This accounts for the result, seen above, and often repeated, that the slope of the search function in the target-absent cases is approximately twice that of target-present cases. Using this logic, the speed of search scanning may be estimated from the slope of the target-absent data and is about 60 ms/item.

Plots similar to those of Figure 5.3 have been produced many times. Figure 5.4 shows a recent one from Motter and Belky (1998a). The left-hand plot shows a standard search function, with the triangles representing feature searches and the circles conjunction searches. The participants in this study were trained rhesus monkeys. The displays contained a target on every trial and the search time represents the time for the animals to 'find' the target with their eyes. The right hand

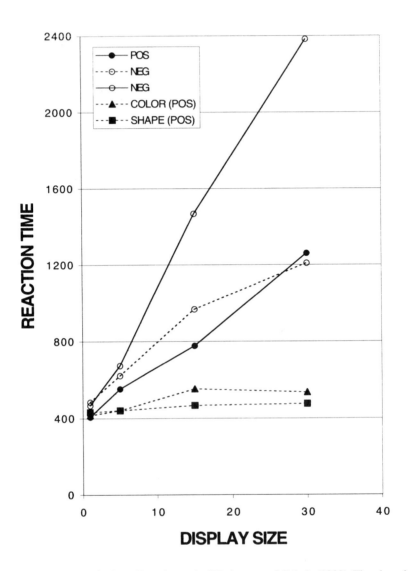

FIGURE 5.3. Results from Experiment 1 of Treisman and Gelade (1980). The plots show a series of search functions, in which the time taken to respond whether a target is present or absent in the display is plotted against the number of elements in the display. The dashed lines (disjunction) show the results for cases where the target can be identified through a single feature (colour or shape), and the solid lines (conjunction) show the results from a feature-conjunction search.

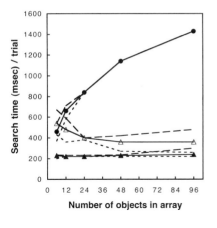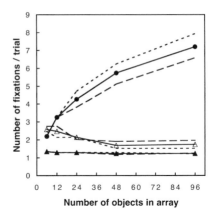

FIGURE 5.4. Results from the experiment of Motter and Belky (1998a) in which monkeys carried out a simple search task (triangle data points) or a conjunction search task looking for a target with a particular colour and orientation (circle data points). The left-hand plot shows the search function, plotting the average time taken to move the eyes to the target against the size of the display. The right-hand plot shows the average number of fixations before the target was located.

plot shows the mean number of fixations before the target was located in this way. The search function for the conjunction task is very comparable with those found when humans carry out a similar task, although the search rate is somewhat faster.

It is clear that the variation in the number of eye fixations is highly correlated with the variation in search times. Similar correlations have been shown in human search (Binello, Mannan and Ruddock, 1995; Zelinsky and Sheinberg, 1997). Conjunction search thus involves serial overt scanning with the eyes. Invoking the assumption discussed earlier about self-terminating search, the data imply that a number of items (approx. 7-8) are dealt with on each fixation to give a search rate of 20 msec/item.

Are the items processed within each fixation dealt with in parallel, or is there a process of covert attentional scanning that scans each item serially? Several pieces of evidence suggest the former. First, both theoretical and empirical work in the Treisman tradition have supported parallel processing of a small number of items (Pashler, 1987; Eckstein, 1998) and Treisman herself has been ready to support this (Treisman and Sato, 1990). Second, more detailed studies of eye movements in search provide support for the same conclusion.

In our laboratory, we have studied a task of feature-conjunction search using displays of the type shown in Fig. 5.5. Subjects were required to move their eyes from the central position to locate a prespecified target. In displays with the target in the inner ring, three out of four subjects tested showed high accuracy in making a single saccade to the target (Findlay, 1997). The latency of these saccades was in the range 200-250 msec, not significantly longer than situations that involved

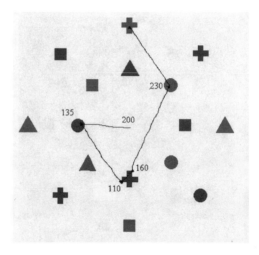

FIGURE 5.5. Feature-conjunction search task of the type studied by Findlay (1997), together with a record of a subject's eye-scan data. The target in this display was a cross located in the centre-top of the display, and subjects were instructed to move their eyes to the target as rapidly as possible. The figures show the fixation times in milliseconds for fixations before the target was reached.

a search for a simple pop-out feature. If the target was in the outer ring, only on a small percentage of occasions was the first saccade directed to it. On the remainder, the saccade was nearly always directed to a distractor in the inner ring. The latency of these erroneous saccades was slightly shorter (218 msec) than those of correct saccades (228 msec). Erroneous saccades were more likely to be directed to distractors sharing a feature with the target than onto distractors sharing neither target feature.

All of these findings are readily explained on the assumption that a number of display elements are processed in parallel during the first fixation but are less easy to explain with a model in which covert attention scans around the display prior to any overt eye movement. In particular, such a serial scanning model would appear to predict that the latencies for incorrect saccades would be longer than those for correct saccades, on the basis of any reasonable hypothesis linking covert to overt scanning. Motter and Belky (1998a,b) also found that the latency of saccades that led to immediate target capture was no different from that in the remainder of cases and used this fact to argue against serial covert scanning. This parallel-processing model also accords with current work on the neurophysiology of visual search (Schall, 1995; Schall and Hanes, 1998; Schall and Thompson, 1999).

The tasks we have described previously were characterised by peripheral-preview effects. There is some indication in visual search that peripheral preview plays a role. This is shown by the occurrence of extremely short fixations (<100 msec) found when eye-scanning records during search are analysed. We have found tat these short fixations occur regularly (Findlay, Gilchrist, and Brown, 1998, 2000) and they have also been reported by McPeek, Keller, and Nakayama

(1999). Indications that peripheral preview is involved come from the finding that the short fixations only occur when the eyes land on (or near) a distractor, never when the eye lands on a target. Occasions when the eye lands on or near a target are often followed by a further saccade, usually a small corrective movement. However, these saccades are preceded by a fixation of normal length (100-180 msec). Because the very brief fixations are too short for visual information to be taken in during the fixation, the decision to make the second saccade must have been already made before the first movement occurred; that is, a pair of "pipelined" saccades is programmed. Because the decision is dependent on the nature of the material at the saccade destination, peripheral preview is involved[5].

Work in the area of visual search thus produces convergent findings with those previously discussed. During a fixation, visual information is processed in parallel with enhanced processing at the area of the following saccade location. There is no support for sequential intra-fixational scanning by covert attention.

5.6   Rethinking Covert Attention

Whenever an eye movement directs the gaze to a location in the visual field, the enhanced visual resolution found in foveal vision becomes available to analyse material at this new location. If attention is moved covertly, this resolution benefit does not operate. Can any complementary benefits be found for covert shifts of attention?

One suggestion might be that the locus of covert attention could be moved around the visual scene more rapidly than the eyes. We have noted cases where researchers have looked for evidence of a fast rate of covert scanning and found no evidence to support such a phenomenon. Further estimates of the rate at which covert attention can be moved between items on a display have been reviewed by Egeth and Yantis (1997). A wide variety of estimates are described, ranging from a few tens of milliseconds per item to several hundred milliseconds per item. Some of the faster estimates come from the slope of search functions and in the previous section we have already argued that alternative interpretations of these data may be made. Other fast estimates (Saarinen and Jules, 1991) come from rapid serial visual presentation (RSVP). When a set of digits are exposed sequentially, but extremely rapidly, the order of the digits can often be reported. This may be used to argue that exogenously driven attention can be moved rapidly (although Egeth and Yantis note certain caveats). Theeuwes, Kramer, Hahn and Irwin (1998) have also produced data showing that exogenous factors may operate during the preparatory stage of an eye movement, but at a location different from

---

[5]Peripheral preview might be involved in two ways. First, it is possible that information from the saccade destination is acquired at a point in time during the programming of the first saccade too late to cancel it but in time to produce a saccade pair. The second possibility is that the peripheral preview of the *target* prevents a second saccade from occurring that would otherwise have been made. Further work will be necessary to distinguish between these possibilities.

that of the saccade destination. During the latency period whilst an observer was preparing a saccade to one target destination, a new visual stimulus was presented at a different location to the saccade target. Frequently this distractor 'captured' the saccade, which was directed to the distractor rather than to the intended target. However, on those occasions in which the eyes were not so captured, improved visual discrimination was found at the location of the distractor.

In the case of endogenous movements of covert attention, estimates of the speed of redeployment are much slower. Duncan, Ward, and Shapiro (1994) and Sperling and Weichselgartner (1995) used variants of the RSVP technique to assess the speed with which endogenous (centrally driven) attention can be relocated. Their estimates (300 - 500 msec) are very much slower and comparable with the speed at which overt eye movements can be made.

We thus argue that there is no convincing evidence that covert attention can move rapidly enough to produce sequential intrafixational scanning unless fixations are artificially prolonged. Covert attention appears to operate in close conjunction with actual saccadic eye movements by preprocessing the material at the location of the forthcoming movement in a way that subsequently allows faster processing of the material when fixated. In the case of reading, this speeds up the rate of information intake by some 10-15%. In this task at least, the absence of covert attention would only impair task performance to a small extent. The loss of the ability to make eye-scanning movements, however, would be devastating.[6]

5.7   Conclusion

We have examined three areas in which visual tasks are carried out using eye s-canning. In reading, scene viewing and free visual search, the eyes make saccadic movements several times each second. We have argued that, in these tasks, there is no evidence for additional spatial scanning by movements of a covert attention spotlight. However, covert attention is involved in these tasks, as shown by the asymmetry of the reading perceptual span and the phenomenon of peripheral pre-view. Nevertheless, the overt movements of the eyes are primary and the covert processes only operate in association with these movements.

References

Aloimonos, J., Bandopadhay, A. and Weiss, I. (1988). Active vision. *Int. J. Comp. Vis.*, 1:333-356.

---

[6]The eye scanning need not necessarily be mediated by eye movements of the conventional type however. We have studied an individual with congenital ophthalmoplegia (Gilchrist, Brown and Findlay, 1997; Gilchrist, Brown and Findlay, & Clarke 1998). She has no functional eye muscles but can nevertheless use her vision to read and carry out other tasks. Saccade-like movements of the head substitute for a normal individual's eye movements.

Anstis, S. M. (1974). A chart demonstrating variations in acuity with retinal position. *Vis. Res.*, 14:589-592.

Ballard, D. H. (1991). Animate vision. *Artif. Intel.*, 48:57-86.

Ballard, D. H., Hayhoe, M. M., Pook, P. K. and Rao, R. P. N. (1998). Deictic codes for the embodiment of cognition. *Behav. Brain Sci.* 20:66-80.

Binello, A., Mannan, S. and Ruddock, K. H. (1995). The characteristics of eye movements made during visual search with multi-element stimuli. *Spatial Vis.*, 9:343-362.

Blanchard, H. E., McConkie, G. W., Zola, D. and Wolverton, G. S. (1984). Time course of visual information utilization during fixations in reading. *J. Exp. Psych. Human Percept. Perf.*, 10:75-89.

Cave, K. R. and Bichot, N. P. (1999). Visuospatial attention: beyond the spotlight model. *Psychonomic Bull. and Rev.*, 6:204-223.

Churchland, P. S., Ramachandran, V. S. and Sejnowski, T. J. (1994). A critique of pure vision. In C. Koch and J. L. Davis (Eds.) *Large Scale Neuronal Theories of the Brain*, pp. 23-60, Cambridge, MA: MIT Press.

Deubel, H. and Schneider, W. X. (1996). Saccade target selection and object recognition: Evidence for a common attentional mechanism. *Vis. Res.*, 36:1827-1837.

Duncan, J., Ward, R. and Shapiro, K. (1994). Direct measurement of attention dwell time in human vision, *Nature*, 369:313-315.

Eckstein, M. P. (1998). The lower visual search efficiency for conjunctions is due to noise and not serial attentional processing. *Psychol. Sci.*, 9:111-118.

Egeth, H. E. and Yantis, S. (1997). Visual attention: control, representation, and time course. *Annu. Rev. Psych.*, 48:269-297.

Findlay, J. M. (1997). Saccade target selection in visual search. *Vis. Res.*, 37:617-631.

Findlay, J. M., Gilchrist, I. D. and Brown, V. (1998). Saccades in search: the role of trans-saccadic memory. *Invest. Ophth. and Vis. Sci.*, 39:S165.

Findlay, J. M. and Walker, R. (1999). A model of saccadic eye movement generation based on parallel processing and competitive inhibition. *Behav. and Brain Sci.*, 22:661-674.

Gibson, J. J. (1966). *The Senses Considered as Perceptual Systems*. Boston: Houghton Mifflin.

Gilchrist, I. D., Brown, V., Findlay, J. M. and Clarke, M. P. (1998). Using the eye movement system to control the head. *Proc. Roy. Soc. B*, 265:1831-1836.

Gough, P. (1972). One second of reading. In J. F. Kavanagh and I. G. Mattingley (Eds.) *Language by Ear and by Eye. The Relationship Between Speech and Reading*. pp. 335-358, Cambridge, MA: MIT Press.

Grimes, J. (1996). On the failure to detect changes in scenes across saccades. In K. Akins (Ed.) *Perception*. Volume 5 of the Vancouver Studies in Cognitive Science. New York, NY: Oxford University Press.

Henderson, J. M. and Ferreira, F. (1990). Effects of foveal processing difficulty on the perceptual span in reading: implications for attention and eye movement control. *J. Exp. Psych.: Learning, Memory and Cognition*, 16:417-429.

Henderson, J. M. and Ferreira, F. (1993). Eye movement control during reading: fixational measures reflect foveal but not parafoveal processing difficulty. *Canadian J. Exp. Psych. - Revue Canadienne de Psychologie Expérimentale*, 47:201-221.

Henderson, J. M. and Hollingworth, A. (1998). Eye movements during scene viewing: an overview. In G Underwood (Ed.) *Eye Guidance in Reading and Scene Perception*. pp. 269-293, Amsterdam: North-Holland/Elsevier.

Henderson, J. M. and Hollingworth, A. (1999). High-level scene perception. *Ann. Rev. Psych.*, 50:243-271.

Henderson, J. M., Pollatsek, A. and Rayner, K. (1989). Covert visual attention and extrafoveal information use during object identification. *Percept. and Psychophys.*, 45:196-208.

Hoffman, J. E. and Subramaniam, B. (1995). The role of attention in saccadic eye movements. *Percept. and Psychophys.*, 57:787-795.

Inhoff, A. W. and Rayner, K. (1986). Parafoveal word processing during eye fixations in reading: effect of word frequency. *Percept. and Psychophys.*, 40:431-439.

Kennedy, A. (1998). The influence of parafoveal words on foveal inspection time: evidence for a processing trade-off. In G. Underwood (Ed.) *Eye Guidance in Reading and Scene Perception*. pp. 149-179. Elsevier: Amsterdam.

Kennison, S. M. and Clifton, C. (1995). Determinants of parafoveal preview benefit in high and low working memory capacity readers: implications for eye movement control. *J. Exp. Psych.: Learning, Memory and Cognition*, 21:68-81.

Kingstone, A. and Klein, R. M. (1993). Visual offset facilitates saccade latency: does pre-disengagement of attention mediate this gap effect? *J. Exp. Psych., Human Percept. and Perf.*, 19:251-265.

Kowler, E., Anderson, E., Dosher, B. and Blaser, E. (1995). The role of attention in the programming of saccades. *Vis. Res.*, 35:1897-1916.

Land, M. F. (1995). The functions of eye movements in animals remote from man. In J. M. Findlay, R. Walker and R. W. Kentridge (Eds.) *Eye Movement Research: Mechanisms, Processes and Applications*. pp. 63-76. Elsevier: Amsterdam.

Land, M. F., Mennie, N. and Rusted, J. (1999). The roles of vision and eye movements in the control of activities of everyday living. *Percept.*, 28: 1311-1328.

Liversedge, S. P. and Findlay, J. M. (2000). Saccadic eye movements and cognitive science. *Trends in Cog. Sci.*, 4:6-14.

Loftus, G. R. and Mackworth, N. H. (1978). Cognitive determinants of fixation location during picture viewing. *J. Exp. Psych., Human Percept. and Perf.*, 4:565-572.

McClelland, J. L. and Mozer, M. C. (1986). Perceptual interactions in two-word displays: familiarity and similarity effects. *J. Exp. Psych., Human Percept. and Perf.*, 12:18-35.

McConkie, G. W. and Rayner, K. (1975). The span of the effective stimulus during a fixation in reading. *Percept. and Psychophys.*, 17:578-586.

McPeek, R. M., Keller, E. L. and Nakayama, K. (1999). Concurrent processing of saccades. *Behav. and Brain Sci.*, 22:691-692.

Marr, D. (1982). *Vision*. San Francisco, CA: W. H. Freeman.

Morrison, R. E. (1984). Manipulation of stimulus onset delay in reading: evidence for parallel programming of saccades. *J. Exp. Psych., Human Percept. and Perf.*, 5:667-682.

Motter, B. C. and Belky, E. J. (1998a). The zone of focal attention during active visual search. *Vis. Res.*, 38:1007-1022.

Motter, B. C. and Belky, E. J. (1998b). The guidance of eye movements during active visual search. *Vis. Res.*, 38:1805-1818.

Müller, H. J. and Rabbitt, P. M. A. (1989). Reflexive and voluntary orienting of visual attention: time course of activation and resistance to interruption. *J. Exp. Psych., Human Percept. and Perf.*, 15:315-330.

Nakayama, K. (1992). The iconic bottleneck and the tenuous link between early visual processing and perception. In C. Blakemore (Ed) *Vision: Coding and Efficiency*, Cambridge University Press.

Nakayama, K. and Mackeben, M. (1989). Sustained and transient components of focal visual attention. *Vis. Res.*, 29: 1631-1647.

Nelson, W. W., and Loftus, G. R. (1980). The functional visual field during picture viewing. *J. Exp. Psych., Hum. Learn. Mem.*, 6:391-399.

O'Regan, J. K. (1992). Solving the 'real' mysteries of visual perception: the world as outside memory. *Can. J. Psych.*, 46:461-488.

Pashler, H. (1987). Detecting conjunction of color and form: re-assessing the serial search hypothesis. *Percept. and Psychophys.*, 41:191-201.

Pashler, H. (Ed.) (1998) *Attention*. Hove: Psychology Press.

Pollatsek, A., Rayner, K. and Collins, W. E. (1984). Integrating pictorial information across saccadic eye movements. *J. Exp. Psych., Gen.*, 113:426-442.

Posner, M. I. (1980). Orienting of attention. *Q. J. Exp. Psych.*, 32:3-25.

Posner, M. I., Nissen, M. J. and Ogden, M. C. (1978). Attended and unattended processing modes: the role of set for spatial location. In H. L. Pick and I. J. Saltzman (Eds), *Modes of Perceiving and Processing Information.* pp. 137-157. Hillsdale: Lawrence Erlbaum.

Posner, M. I., Walker, J. A., Friedrich, F. J. and Rafal, R. D. (1984). Effects of parietal lobe injury on covert orienting of visual attention. *J. Neurosci.*, 4:1863-1874.

Rayner, K. (1984). Visual search in reading, picture perception and visual search. A tutorial review. In H. Bouma and D. Bouwhuis (Eds.) *Attention and Performance X*. pp. 67-96. Hillsdale, NJ: Lawrence Erlbaum.

Rayner, K. (1998). Eye movements in reading and information processing. 20 years of research. *Psychol. Bull.*, 124:372-422.

Reichle, E. D., Pollatsek, A., Fisher, D. F. and Rayner, K. (1998). Toward a model of eye movement control in reading. *Psychol. Rev.*, 105:125-147.

Rensink, R., O'Regan, J. K. and Clark, J. J. (1997). To see or not to see: the need for attention to perceive changes in scenes. *Psychol. Sci.*, 8:368-373.

Rensink, R. A. (2000). Seeing, sensing, and scrutinizing. *Vis. Res.*, in press.

Rizzolatti, G., Riggio, L., DaScola, I. and Umiltá, C. (1987). Reorienting attention across the horizontal and vertical meridians: evidence in favor of a premotor theory of attention. *Neuropsychologia*, 25:31-40.

Rizzolatti, G., Riggio, L. and Sheliga, B. M. (1994). Space and selective attention. In C. Umiltá and M. Moscovitch (Eds.) *Attention and Performance XV*. pp 231-265. Cambridge, MA: MIT Press.

Rovamo, J., Virsu, V. and Näsänen, R. (1978). Cortical magnification factor predicts the photopic contrast sensitivity of peripheral vision. *Nature*, 271:54-55.

Saarinen, J. and Julesz, B. (1991). The speed of attentional shifts in the visual field. *Proc. Nat. Acad. Sci.*, 88:1812-1814.

Saida, S. and Ikeda, M. (1979). Useful visual field size for pattern perception. *Percept. and Psychophys.*, 25:119-125.

Schall, J. D. (1995). Neural basis of saccade target selection. *Rev. Neurosci.*, 6:63-85.

Schall, J. D. and Hanes, D. P. (1998). Neural mechanisms of selection and control of visually guided eye movements. *Neural Networks*, 11:1241-1251.

Schall, J. D. and Thompson, K. G. (1999) Neural selection and control of visually guided eye movements. *Ann. Rev. Neurosci.*, 22:241-259.

Shiori, S. and Ikeda, M. (1989). Useful resolution for picture perception as a function of eccentricity. *Percept.*, 18:347-361.

Schroyens, W., Vitu, F., Brysbaert, M. and d'Ydewalle, G. (1999). Eye movement control during reading: foveal load and parafoveal processing. *Q. J. Exp. Psych.*, 52A:1021-1046.

Shepherd, M., Findlay, J. M. and Hockey, G. R. J. (1986). The relationship between eye movements and spatial attention. *Q. J. Exp. Psych.*, 38A:475-491.

Slowiaczek, M. L. and Rayner, K. (1987). Sequential masking during eye fixations in reading. *Bul. Psychonom. Soc.*, 25:175-178.

Sperling, G. S. and Weichselgartner, E. (1995). Episodic theory of the dynamics of spatial attention. *Psych. Rev.*, 102:503-532.

Styles, E. (1997). *The psychology of attention.* Hove: Psychology Press.

Theeuwes, J., Kramer, A. F., Hahn, S. and Irwin, D. E. (1998). Our eyes do not always go where we want them to go. *Psychol. Sci.*, 9:379-385.

Treisman, A. and Gelade, G. (1980). A feature integration theory of attention. *Cog. Psych.*, 12:97-136.

Treisman, A. and Sato, S. (1990). Conjunction search revisited. *J. Exp. Psych., Human Percept. and Perf.*, 16:459-478.

Van Diepen, P. M. J., Wampers, M. and d'Ydewalle, G. (1998). Functional division of the visual field: moving masks and moving windows. In G. Underwood (Ed.) *Eye Guidance in Reading and Scene Perception.* Chapter 15, pp. 337-355. Amsterdam: Elsevier.

Wertheim, T. (1894). Über die indirekte Sehschärfe. Z. Psych. 01 Phys. Sinnesorgans, 7:121-187.

Wright, R. D. (Ed) (1998). *Visual Attention.* New York, NY: Oxford University Press.

Yeshurun, Y. and Carrasco, M. (1998). Attention improves or impairs visual performance by enhancing spatial resolution. *Nature*, 395:72-75.

Zelinsky, G. J. and Sheinberg, D. L. (1997). Eye movements during parallel-serial visual search. *J. Exp. Psych. Human Percept. and Perf.*, 23:244-262.

6

Complexity, Vision, and Attention

John K. Tsotsos

What does it mean for a problem to be *complex*? One dimension of complexity is *computational complexity*. This chapter focuses on how this type of complexity affects the design of perceptual systems. Many natural problems have optimal solutions that are believed to be computationally intractable in any implementation, machine or neural. Thus, the computational complexity of a particular solution greatly affects its realizability, and thus its plausibility. The focus here will be on problems in vision.

This chapter provides:

- a brief introduction to the theory of computational complexity;

- an explanation of why this discrete theory is applicable to the study of biological systems;

- several examples of formal problems within vision that are believed to be intractable;

- guidelines for how to deal with intractable problems;

- an application of those guidelines to vision;

- an overview of a model of visual attention that results from that application.

6.1   What Is Computational Complexity?

Computational complexity is studied to determine the intrinsic difficulty of mathematically posed problems that arise in many disciplines.[1] Many of these problems involve combinatorial search; (i.e., search through a finite but extremely large, structured set of possible solutions). Examples include the placement and interconnection of components on an integrated-circuit chip, the scheduling of major league sports events, or bus routing. Complexity theory tries to discover the limitations and possibilities inherent in a problem. In the same way that the laws of

---

[1]See Garey and Johnson (1979) and Stockmeyer and Chandra (1979) for further discussion and background.

thermodynamics provide theoretical limits on the utility and function of nuclear power plants, complexity theory provides theoretical limits on information processing systems. If biological vision can indeed be computationally modeled, then complexity theory is a natural tool for investigating the information processing characteristics of both computational and biological vision systems.

Using complexity theory, one can ask: For a given computational problem, how well, or at what cost, can it be solved? Before studying complexity one must define an appropriate complexity measure. Several measures are possible, but the common ones are related to the space requirements (numbers of memory or processor elements) and time requirements (how long it takes to execute) for solving a problem.

In what sense are complexity results inherent to a particular problem? Certain intrinsic properties of the universe will always limit the size and speed of computers. Consider the following argument from Stockmeyer and Chandra (1988): The most powerful computer that could conceivably be built could not be larger than the known universe (less than 100 billion light-years in diameter), could not consist of hardware smaller than the proton (10^{13} cm in diameter), and could not transmit information faster than the speed of light (3 x 10^8 m/sec). Given these limitations, such a computer could consist of at most 10^{126} pieces of hardware. It can be proved that, regardless of the ingenuity of its design and the sophistication of its program, this ideal computer would take at least 20 billion years to solve certain mathematical problems that are known to be solvable in principle. Because the universe is probably less than 20 billion years old, it seems safe to say that such problems defy computer analysis.

6.1.1 Some basic definitions

The following are some basic definitions common in complexity theory (Garey and Johnson, 1979). A *problem* is a general question to be answered, usually possessing several parameters whose values are left unspecified. A problem is described by giving a general description of all of its parameters and a statement of what properties the answer, or *solution*, is required to satisfy. An *instance* of the problem is obtained by specifying particular values for all of the problem parameters. An *algorithm* is a general step-by-step procedure for achieving solutions to problems. To solve a problem means that an algorithm can be applied to any problem instance and is guaranteed to always produce a solution for that instance.

The time requirements of an algorithm are expressed in terms of a single variable, n, reflecting the amount of input data needed to describe a problem instance. A *time complexity function* for an algorithm expresses its time requirements by giving, for each possible input length, an upper bound on the time needed to achieve a solution. If the number of operations required to solve a problem is an exponential function of n, then the problem has *exponential time complexity*. If the number of required operations can be represented by a polynomial function in n, then the problem has *polynomial time complexity*. Similarly, *space complexity* is defined as a function for an algorithm that expresses its space or memory requirements. *Algorithmic complexity* is the cost of a particular algorithm. This should be contrasted with *problem complexity*, which is the minimal cost over all possible algorithms. The dominant kind of analysis is *worst-case*: at least one instance out of all possible instances has this complexity.

A worst-case analysis provides an upper bound on the amount of computation that must be performed as a function of problem size. If one knows the maximum problem size, then the analysis places an upper bound on computation for the whole problem as well. Thus, one may then claim, given an appropriate implementation of the problem solution, that processors must run at a speed dependent on this maximum in order to ensure real-time performance for all possible inputs. Worst cases do not only occur for the largest possible problem size; rather, the worst-case time-complexity function for a problem gives the worst case number of computations for any problem size; this worst-case may be required simply because of unfortunate ordering of computations (for example, a linear search through a list of items would take a worst-case number of comparisons if the item sought is the last one). Thus, worst-case situations in the real world may happen frequently for any given problem size.

Critical ideas in complexity theory are that of *complexity class* and, related to it, *reducibility*. If a problem S is known to be efficiently transformed (or reduced) to a problem Q, then the complexity of S cannot be much more than the complexity of Q. *Efficiently reduced* means that the algorithm that performs the transformation has polynomial complexity. The *class* P consists of all those problems that can be solved in polynomial time. If we accept the premise that a computational problem is not tractable unless there is a polynomial-time algorithm to solve it, then all tractable problems belong in P.

In addition to the class P of tractable problems, there is also a major class of presumably intractable problems. If a problem is in the class NP, then there exists a polynomial $p(n)$ such that the problem can be solved by an algorithm having time complexity of the order of $2^{p(n)}$; that is, the time complexity function is asymptotically (as n becomes large) dominated by the polynomial $p(n)$. A problem is *NP-complete* if it is in the class NP, and it polynomially reduces to an already proven NP-complete problem. The set of such problems form an equivalence class. Clearly, there must have been a first NP-complete problem. The first such problem was that of satisfiability (Cook's 1971 Theorem). There are hundreds of NP-complete problems. If any NP-complete problem can be solved in polynomial time, then they all can. Most doubt the possibility that nonexponential algorithms for these problems will ever be found, so proving a problem to be NP-complete is now regarded as strong evidence that the problem is intrinsically intractable. If an efficient algorithm can be found for any one (and hence all) NP-complete problems, however, it would be a major intellectual breakthrough.

6.1.2   Dealing with NP-completeness

NP-completeness effectively eliminates the possibility of developing a completely optimal and general algorithm. Once a problem is seen to be NP-complete, it is appropriate to direct efforts toward a more achievable goal. In most cases, a direct understanding of the size of the problems of interest and the size of the processing machinery is of tremendous help in determining which are the appropriate approximations. A variety of approaches have been taken when confronted with an NP-complete problem:

1. Develop an algorithm that is fast enough for small problems but that would take too long with larger problems. This approach is often used when the anticipated problems are small.

2. Develop a fast algorithm that solves a special case of the problem but does not solve the general problem. This approach is often used when the special case is of practical importance.

3. Develop an algorithm that quickly solves a large proportion of the cases that come up in practice but in the worst case may run for a long time. This approach is often used when the problems occurring in practice tend to have special features that can be exploited to speed up the computation.

4. For an optimization problem, develop an algorithm that always runs quickly but produces an answer that is not necessarily optimal. Sometimes, a worst-case bound can be obtained on how much the answer produced may differ from the optimum, so that a reasonably close answer is assured. This is an area of active research, with suboptimal algorithms for a variety of important problems being developed and analyzed.

5. Use natural parameters to guide the search for approximate algorithms. There are a number of ways a problem can be exponential. Consider the natural parameters of a problem rather than a constructed problem length, and attempt to reduce the exponential effect of the largest-valued parameters.

6.1.3   Vision and NP-completeness

Are there any vision problems that are provably in the class of NP-complete problems? There are several:

1. Unbounded visual search using a passive sensor system is NP-complete (Tsotsos, 1989, 1990);

2. Unbounded visual search using an active sensor system is NP-complete (Tsotsos, 1992);

3. Polyhedral scene line-labelling is NP-complete (Kirousis and Papadimitriou, 1988);

4. Relaxation procedures for constraint-satisfaction networks are P-Complete[2] (Kasif 1990);

5. Finding a single, valid interpretation of a scene with occlusion is NP-hard[3] (Cooper, 1992);

6. Curved object line labelling is NP-complete (Dendris, Kalafatis, and Kirousis 1994);

7. Unbounded stimulus-behavior search is NP-hard (Tsotsos, 1995);

8. 3-D sensor planning is NP-complete (Ye and Tsotsos, 1996).

It is probably true that most "general" problems dealing with perception are intractable. Using neural networks does not magically provide the answer; Judd proved a wide variety of connectionist problems to be intractable (starting with the loading problem; Judd 1990). Assuming $P \neq NP$, these problems cannot be solved in their general form with realizable hardware in reasonable amounts of time, and it does not matter whether the implementation is neural or silicon-based. Many use human vision as the benchmark against which one measures "general purpose" vision capabilities. But human vision cannot be solving the general problem! (Tsotsos 1990).

The previous listing only scratches the surface of the literature on the topic; there are many more examples, and they form quite broad and natural problem classes. It appears that any interesting problem related to human intelligence has the characteristic that it is susceptible to combinatorial explosion.

6.2   Can Perception Be Modeled Computationally?

All of the previous discussion and theory is relevant to perception only if it can be shown that perception can be modeled computationally. A proof of *decidability* is sufficient to guarantee that a problem can be modeled computationally (see Davis 1958 and 1965, for in-depth discussions of decidability).[4] If it is the case that for some problem we wish to know of each element in a countably infinite set A, whether or not that element belongs to a certain set B, which is a proper subset of A, then that problem can be formulated as a decision problem. Such a

---

[2] P-Complete is the class of problems for which no efficient parallel algorithms can be found; that is, those problems have components that are inherently sequential. This problem is included in this list for two reasons. First, the relaxation procedure itself is very common in vision and neural network models; it is used in line-labelling algorithms for example. Second, in general, it may be that perception is inherently sequential and this proof is one piece of evidence towards this conclusion.

[3] Problems that are NP-hard are as expensive to solve as those that are NP-complete; the NP-complete term, however, is reserved for decision problems whereas NP-hard is used for all others.

[4] Decidability should not be confused with tractability. An intractable problem may be decidable; but for an undecidable problem, one cannot determine its tractability.

problem is decidable if there exists a *Turing Machine*[5] which computes yes or no for each element of A. This requires that perception, in general, be formulated as a decision problem. This formulation does not currently exist. If no subproblem of perception can be found to be decidable, then it might be that perception as a whole is undecidable and thus cannot be computationally modeled. But, at least one decidable perceptual problem does exist. Visual search, an important subproblem, can be formulated as a decision problem (Tsotsos, 1989) and is decidable; it is an instance of the Comparing Turing Machine defined in Yashuhara 1971. More research is needed to try to formalize other sub-problems of perception in the same way; note that all of the problems listed in Section 6.3 are also decidable ones even if they are intractable. Even if some other aspect of perception is determined to be undecidable, this does not mean that all of perception is also undecidable nor that other aspects of perception cannot be modeled computationally. For example, one of the most famous undecidable problems is whether or not an arbitrary Diophantine equation has integral solutions (Hilbert's 10th problem). This does not mean that mathematics cannot be modeled computationally! Similarly, another famous undecidable problem is the halting problem for Turing Machines: it is undecidable whether a given Turing Machine will halt for a given initial specification of its tape. This, too, has important theoretical implications, but because Turing Machines form the foundation of computation, it certainly does not mean that computation cannot exist!

Any computational paradigm is a candidate for use in constructing a biologically plausible model. Neural network approaches are not the only ones that are biologically plausible as is often believed. Neural networks are Turing-equivalent and are subject to the same constraints of computational complexity and computational theory as any other implementation (see Judd, 1990, for further discussion and proofs of this statement). It is important to note that relaxation processes are specific solutions to search problems in large parameter spaces and nothing more. Neural networks use variations of such search procedures which in general may be termed optimization techniques. If optimization is the process by which real neurones perform some of their computation, it is subject to precisely the same considerations of computational complexity as any other search scheme.

Many argue that worst-case analysis is inappropriate for perception. Some say that relying on worst-case analysis and drawing the link to biological vision implies that biological vision handles the worst-case scenarios. This kind of inference is quite incorrect. As was shown in Tsotsos (1990), it is impossible for the biological (or any other) visual system to handle worst-case scenarios. The whole argument exists only to prove that all worst-case scenarios cannot be handled by human vision in a bottom-up fashion and that the quest for general solutions is futile. It has also been said that biological vision systems are designed around average or perhaps

---

[5]A Turing Machine is a hypothetical computing device (Turing 1937). A Deterministic Turing Machine consists of a finite state control, a read-write head, and a two-way infinite sequence of labelled tape squares. A program then provides input to the machine, is executed by the finite state control, and computations specified by the program read and write symbols on the squares of the tape.

best-case assumptions. However, it is far from obvious what kind of assumptions (if any) went into the design of biological vision systems. Vision systems emerged as a result of a complex interaction of many factors, including a changing environment, random genetic mutations, and competitive behavior. It is probably the case that the best we will ever be able to do under such circumstances is to place an upper bound on the complexity of the problem, and this is all worst case analysis will provide. Finally, many say that expected case analysis more correctly reflects the world that biological vision systems see. Analyses based on expected or average cases, depend critically on having a well-circumscribed domain and an algorithm. Thus, the complexity measures derived reflect algorithmic complexity and not problem complexity. Only under those conditions can average- or expected-case analyses be performed. In general, it is not possible to define what the average or expected input is for a vision system in the world. Furthermore, the result of the analysis will be valid only for the average input and does not place a bound on the complexity of the vision process as a whole. It is important to note, however, that Parodi et al. (1998) have shown that with an average case analysis, a provably exponential vision task (polyhedral scene line labelling) becomes linear with the addition of a small amount of task-specific knowledge (label ordering). Thus, even in the average case, our basic result seems to hold.

Complexity theory is as appropriate for analysis of perception in general as any other analysis tool currently used by biological experimentalists. Experimental scientists attempt to explain their data and not just describe it; it is no surprise that their explanations are typically well-thought-out and logically motivated, involving procedural steps or events. In this way, a proposed course of events is hypothesized to be responsible for the data observed. There is no appeal to non-determinism, nor to oracles (nor should there be!) that guess the right answer, nor to undefined, unjustified, or undreamed-of mechanisms that solve difficult components.[6] In essence, experimental scientists attempt to provide an *algorithm* whose behavior leads to the observed data. Attempts at providing algorithmic explanations appeared even before the invention of the computer. For example, Helmholtz's unconscious inference theory (von Helmholtz, 1963) is remarkably similar to the current reasoning paradigm in artificial intelligence, where reasoning is formalized as a logical process using formal mathematics. Whether a proposed algorithm or explanation is realizable depends in part on its tractability.

6.3   Visual Search

As given earlier, visual search is one perceptual problem that has been formalized as a decision problem and is shown to be decidable. Some detail is now given

---

[6]The decidability of perception has far-reaching implications for the development of perceptual theories. One of those implications is that theories that appeal to oracles are probably outside the realm of science.

regarding this formulation.

6.3.1   Definition

Visual search is a common, if not ubiquitous, subtask of vision in both man and machine. A basic visual search task is defined as follows (Rabbitt, 1978): Given a target and a test image, is there an instance of the target in the test image? One may also ask subjects to find "odd-man-out" elements of a display, or simply to describe an image. Typically, experiments measure the time taken to reach a correct response. Region growing, shape matching, structure from motion, the general alignment problem, and connectionist recognition procedures are specialized versions of visual search in that the algorithms must determine which subset of pixels is the correct match to a given prototype or description. The basic visual search task is precisely what any model-based computer vision system has as its goal: Given a target or set of targets (models), is there an instance of a target in the test display? Even basic vision operations such as edge-finding are also in this category: given a model of an edge, is there an instance of this edge in the test image? It is difficult to imagine any vision system that does not involve similar operations. It is clear that these types of operations appear from the earliest levels of vision systems to the highest.

6.3.2   Theory

It was shown in Tsotsos (1989, 1992) that unbounded visual search (no target is given, and even if given, it cannot be used to optimize search), regardless of whether the images are time-varying or the camera system is dynamically controlled (active), is NP-complete. This is due solely to the fact that the subset of pixels in an image that corresponds to a target cannot be predicted in advance and all subsets must be considered in the worst case. The bounded problem (the target is given in advance and is used to optimize search) on the other hand, requires linear time for the search process. This qualitatively confirms all of the visual-search data that have been experimentally discovered (say, by Treisman, 1988) showing a linear response time versus number of items in display relationship. The four theorems proved in Tsotsos (1989, 1992) show that, in general, a bottom-up approach to perception (as suggested by Marr, 1982) is not only computationally intractable, but biologically implausible.

The formulation of the visual-search problem proceeds as follows. There is one objective function to optimize for each known object or image event model. All objective functions are considered in parallel. The search process first seeks the image subsets that satisfy each objective function and, finally, the best match among all of these. The best match is the image subset and model that exhibits the smallest matching error and the model must explain (or cover) as much of the image subset as possible. The brute-force search strategy then is to match each objective function against all possible image subsets. Given formally, the best fit of model to data is sought such that the following is satisfied:

$$\sum_{x \in M, j_x \in I'} |x - j_x| < \theta \text{ and } \sum_{x \in M, j \in I'} x \cdot j_x > \phi \qquad (6.1)$$

The first term is the error measure, while the second is the cover measure. The input is the set I, I' is a subset of I, and M is a set of values corresponding to a particular object or event in the model base. Both I and M are retinotopic representations of the same form. θ and ϕ are two thresholds. I is not necessarily the image itself but may be a collection of all features computed from a given image. A correspondence between elements of M and elements of I' can be hypothesized where element j_x in I' is the element corresponding to x in M. Each possible combination of correspondences may be considered as a separate hypothesis.

Suppose a test image is made up of 256 pixels and a target image has 64 pixels. The correspondence required previously is for each element of the target image (each pixel) to be mapped onto a unique pixel of the test image. This forms a hypothesis about where exactly in the test image the target image is believed to be represented. The spatial organization of the mapping need not preserve the structure of the target stimulus, that is, pixels chosen for the mapping may be arbitrarily distributed throughout the image. Therefore for this example, there are $\binom{256}{64}$ such possible, bottom-up mappings; $\frac{256!}{64!192!}$ is approximately 10^{56}! If spatial structure is preserved and there is no rotation or scaling of the target in the test image, then there are only 81 possibilities such that the target image is entirely within the test image.

Define selective attention as follows. Selective attention is a search-optimization mechanism that tunes the visual processing machinery so that it dynamically approaches an optimal configuration, based on both data-driven and task/knowledge-driven influences. Then, attention selection may determine which mapping to attempt to verify first. If the first such mapping selected is a good one, a great deal of search can be avoided, otherwise there is the potential for a very inefficient search process. For sufficiently small images and/or massive computational power, this brute force concept will work perfectly well. For the brain, this approach fails.

6.3.3 Implications

If the practice of complexity analysis is to lead to tangible benefits then the theorems must lead to algorithms that must be physically realizable, and the physical realization must in some way be better than others with respect to time or space efficiency. No matter what the time and space complexity functions, there is an infinite space of possible variable values or problem sizes that will not be practically realizable. The fact that all computers have finite memories is sufficient to guarantee this. One cannot in practice take infinite time to read or load an infinite Turing Machine tape. Engineering design specifications always impose constraints: the amount of memory may be limited by power consumption or cost; the number of processors is likewise constrained; real-time response places a hard constraint on time complexity and thus on problem size. These constraints cannot be ignored in any complexity discussion that may eventually be used to solve real problems. What are the con- straints whose satisfaction is required in order for a theory to be biologically plausible?

Biological plausibility of a theory of visual perception will be characterized in three stages. First, a theory must be sufficient to explain the observations. Second, it is important to define the size of problem which the algorithm must be able to handle, and this is done as follows:

- The algorithm that embodies the theory is required to accept no more than the same number of input samples of the world per unit time as human sensory organs. It is a non-trivial task to determine exactly the quantitative nature of the input to the human sensory system. With respect to the visual system, there are two eyes; each has about 110–125 million rods and 6.3–6.8 million cones; each eye can discriminate over a luminance span of 10 billion to one; the spatial resolution of the system peaks at about 40 cycles/deg whereas the temporal resolution peaks at about 40 Hz but the two are not independent; finally, there are many inputs from other sensory and motor areas. See Dowling (1987) for further discussion.

- The implementation that realizes the algorithm exists in the real world and requires amounts of physical resources that exist.

- The output behavior of the implementation as a result of those stimuli is comparable both in quality, quantity, and timing to human behavior. The behavioral literature on exactly what the quality, quantity, and timing of human behavior is to a variety of stimuli is immense, but far from complete. What is required, however, are responses from the algorithm that agree qualitatively and quantitatively with human responses and that are generated with the same time delays as human responses. The third stage of the definition requires that the functions for time and space complexity require values of their variables that lead to brain-sized space requirements and behaviorally-confirmed time requirements. Issues of polynomial versus exponential do not enter the discussion of biological plausibility at all. In other words:

- Solutions should require significantly fewer than about 10^9 processors operating in parallel, each able to perform one multiply–add operation over its input per millisecond;

- Processor average fan-in and fan-out should average about one thousand overall; and

- Solutions should not involve more than a few hundred sequential processing steps. Any visual theory must satisfy the preceding characterization, and similarly, any theory of any other aspect of intelligent behavior would have a corresponding characterization of biological plausibility.

6.4   Complexity Level Analysis of Vision

The response times measured by visual-search experiments are assumed to reflect the amounts and kinds of visual-information processing leading to the response. Typically one sees (for many classes of experiment) response time versus input size graphs. Recall the definition of time complexity: time complexity is given as the cost in time as the size of the input set varies. The connection to computational complexity analysis seems rather direct.

It is clear that the brain cannot be solving the intractable form of visual search. It is equally clear that everyday vision is not always directed by a task (i.e., is not always an instance of the bounded form of visual search). It is important to rely not only on the bounded, and linear forms of visual search; the unbounded form must also be carefully considered in order to determine what kind of unbounded problem the brain may be solving. It is thus appropriate to consider the guidelines of Section 6.1.2. Beginning from the NP-completeness of Unbounded Visual Search, complexity-level analysis attempts to follow the guidelines presented in Section 6.1.2 for dealing with such difficult problems.[7] In particular, here guideline 5 is used most effectively in concert with guideline 4.

In Tsotsos (1990), a sequence of modifications to the problem of Unbounded Visual Search is given, driven by the size of the perception problem in terms of number of photoreceptors, feature types, size of visual areas, connectivity of visual areas, for example, in order to transform the problem into a tractable one. The result is a visual search problem that is tractable in time and space (requiring no more processing machinery than the brain may afford) but is not guaranteed to always find optimal solutions. The solutions found are approximate ones; quite acceptable most of the time but sometimes requiring other mechanisms (such as eye movements) or sometimes lacking in precision to some degree. The claim is that this is the form of the visual search problem that the brain is actually solving; a conclusion of this sort is the only possible one because it has been proved that optimal solutions for the general problem of visual search lead to intractability.

The key approximations and optimizations revealed by complexity level analysis are quite straightforward (obvious once laid out in this way!):

- hierarchical processing and organization, which yields a logarithmic improvement in search time within the set of models;

- localized receptive field structure, which reduces the number of image subsets to consider from a function in 2^P to a polynomial of order $P^{1.5}$;

- logically separable feature maps, which permit separate selection of relevant features;

---

[7]Other examples of this can be found in Kirousis (1990, 1993) and Dendris, Kalafatis and Kirousis (1994). Those works begin with intractability proofs for line-labelling problems and develop efficient algorithms for several restricted subproblems.

- visual attention, which permits selection in image space and in feature space for relevant image components and the inhibition of the irrelevant.

Together, these reduce the time and space requirements of the solution to visual search. The first three seem to relate more to hardware organization. Attention, on the other hand, refers to the ability of the system to be tuned in a selective fashion to permit the hardware to be locally optimized for the current task (this view is also detailed in Maunsell, 1995; see also Desimone and Duncan, 1995, for an overview of neuronal mechanisms for visual attention). The remainder of this chapter will highlight aspects of the model of visual attention that resulted from these conclusions.

6.5   The Selective-Tuning Model of Visual Attention

Complexity analysis leads to the conclusion that attention must tune the visual-processing architecture to permit task-directed processing. Selective tuning takes two forms: spatial selection is realized by inhibition of irrelevant connections; and feature selection is realized by inhibition of the units that compute non-selected features. Only a brief summary is presented here (a more detailed account is in Tsotsos et al., 1995). The starting point for the model was described in the formalization of the visual-search problem.

The role of attention in the image domain is to localize the set I' of Equation (6.1) in such a way so that any interfering or corrupting signals are minimized. In doing so, attention also seeks to increase the discriminability of a particular image subset and objective function pair over other such competing pairings as quickly as possible. The search process that localizes the image subset I' is as follows. The visual-processing architecture is assumed to be pyramidal in structure with units within this network receiving both feedforward and feedback connections (the model has this in common with the architecture developed in Van Essen et al. 1992). When a stimulus is first applied to the input layer of the pyramid, it activates in a feed-forward manner all of the units within the pyramid to which it is connected; the result is that an inverted sub-pyramid of units and connections is activated. It is assumed that response strength of units in the network is a measure of goodness-of-match of stimulus to model and of relative importance of the contents of the corresponding receptive field in the scene.

Selection relies on a hierarchy of winner-take-all (WTA) processes. WTA is a parallel algorithm for finding the maximum value in a set. First, a WTA process operates across the entire visual field at the top layer: it computes the global winner (i.e., the units with largest response). The WTA can accept guidance for areas or stimulus qualities to favor if that guidance were available but operates independently otherwise. The search process then proceeds to the lower levels by activating a hierarchy of WTA processes. The global winner activates a WTA that operates only over its direct inputs. This localizes the largest response units within the top-level winning receptive field. Next, all of the connections of the

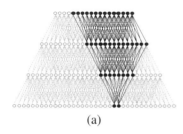 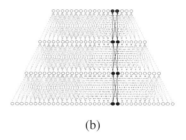

(a)   (b)

FIGURE 6.1. (a) A hypothetical visual processing pyramid showing the portion of the pyramid activated due to the initial feedforward stimuluation of a single input stimulus. (b). The final configuration of the attentional beam reacting to a single input stimulus.

visual pyramid that do not contribute to the winner are pruned. This strategy of finding the winners within successively smaller receptive fields layer by layer in the pyramid and then pruning away irrelevant connections is applied recursively through the pyramid. The end result is that from a globally strongest response, the cause of that largest response is localized in the sensory field at the earliest levels. The paths remaining may be considered the pass zone while the pruned paths form the inhibitory zone of an attentional beam. The WTA does not violate biological connectivity or time constraints.

Figure 6.1 shows a hypothetical visual processing pyramid. There are four layers, each unit connected to seven units in the layer above it and seven units in the layer below it. The input layer (bottom layer) is numbered 1, while the output layer (top layer) is numbered 4. The two examples that follow are intended to illustrate the structure and time course of the application of attentional selection in the model. The first example shows the structure that results if a single stimulus is placed in the visual field. The second example shows the time course of attentional selection if two stimuli are placed in the visual field.

In Fig. 6.1A, only the feedforward connections are shown; the feedback connections are analogous. A stimulus that spans two units in the input layer is to be attended by the system; the resulting beam is shown in Figure 6.1. The light gray lines represent inactive connections, the dotted grey lines represent connections whose feedforward flow is inhibited by the attentional beam, and the black lines represent feedforward connections activated by the stimulus. Black units are activated solely by the black stimulus.

The WTA mechanism locates the peaks in the response in the output layer of the pyramid, shown as the two remaining black units in Figure 6.1B. The inhibitory beam then is extended from top to bottom, pruning away the connections that might interfere with the selected units. Eventually, the two stimulus units are located in the input layer and isolated within the beam.

The important missing link in the preceding example is exactly how the WTA process locates the two winners in the output layer. On the assumption that each of the units in the pyramid computes some quantity using a Gaussian weighted function across its receptive field, then the maximum responses of these computations

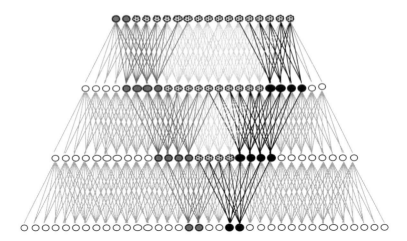

FIGURE 6.2. The visual processing pyramid at the point where the activation due to two separate stimuli in the input layer has just reached the output layer. No attentional effects are yet in evidence.

(whatever they may be) will be exactly the two units selected in the output layer (see Tsotsos et al. 1995, for more detail on this). More importantly, with respect to attention, how does the mechanism function if there is more than one stimulus in the input - that is, with target as well as distractor elements in the visual field?

Figure 6.2 shows the first step of a five-step sequence depicting the changes that the visual processing pyramid undergoes in such a situation. Using the same network configuration as in the previous figure and again showing only the feedforward connections, two stimuli are placed in the visual field (input layer). They are coded black and medium gray as are the connections and units which are activated soley by them. The textured units and light gray connections are those which are activated by both stimuli regardless of proportions. Note that much of the pyramid is affected by both stimuli and as a result, most of the output layer gives a confounded response.

The textured units response is a weak one due to the conflict that arises since each of those units 'sees' two different stimuli within its receptive field. Now the subject is directed to attend to the location of the black stimulus. Location is determined by a mechanism outside the network shown here; appropriate units corresponding to the location in the output layer are marked as shown in Fig. 6.3.

Attention is focused at the black units in the output layer. This is not to say that the output of those units reflects the desired input at this point in time. Rather, these units form the root of the attentional beam as it begins its downward traversal through the visual pyramid. The next phase of the computation is to push the beam down one level further, locating the units that will be the attended ones within the beam.

Simultaneously, the feedforward connections from all units in the layer 3 that feed the attended units in the output layer are inhibited and are not part of the

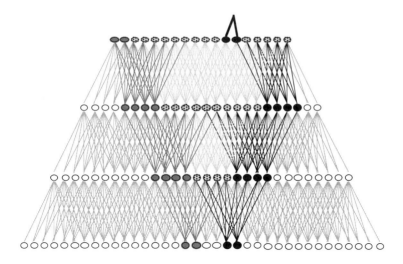

FIGURE 6.3. Attention is focused at the location of the output layer corresponding to the location of the selected input.

pass zone. Figure 6.4 shows the connections whose feedforward flow is explicitly inhibited by the attentional mechanism between layers 3 and 4. The interesting thing to note is that at this early stage of the application of the attentional beam, very little seems to be changing. The large-scale changes come later as more of the visual pyramid is affected by the flow of the attentional beam through it. The selection of units also moves down one level to layer 3.

The next major milestone in this process is shown in Fig. 6.5. At this point of the beam traversal, major changes can be seen. The next set of connections – those between the middle layers – are inhibited. In turn, those inhibitions cause several units in layer 3 to have no active input and thus they provide no signal to the output layer. Those connections are coloured gray, the same as the other inactive connections in the pyramid.

This change, in turn, causes several units in the output layer previously coded textured to be coded black; that is, they receive signals originating only from the black input stimulus. The final stage of the process leads to the network shown in Fig. 6.6. After this point, the selected units in the output layer receive input only from the selected stimulus in the input layer. Note that several units in the output layer are coded in dark gray, showing that the effect of the dark gray stimulus still gets through the beam structure, in fact stronger than in the unattended case of Fig. 6.2.

A few of the weakly responding textured units still remain as well. The interference between stimuli evident in Figure 6.2 is eliminated completely with respect to the item attended and much reduced for the unattended item. The events depicted with this set of figures would occur in the 100 to 200 msecs after stimulus onset (for example, as shown in Chelazzi et al. 1993). Note the difference between the

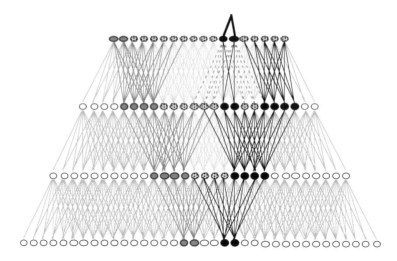

FIGURE 6.4. The first level of inhibition due to the attentional beam. The feedforward flow of the black connections is inhibited.

pattern of activations in Fig. 6.1 and 6.6. In the former case, no location cue is given; the winner-take-all mechanism chooses the strongest responses in the output layer and inhibits the rest. Thus, the set of connections and units attended forms the structure shown with the active connections being strictly those permitted by the selection mechanism. In the latter case, a location cue is given; thus, there is no inhibition within the output layer (if there were, none of the dark gray or textured units would survive).

Suppose one records from one of the textured "neurons" in layer 3 of Fig. 6.5 during the entire process. What will the response over time look like? On stimulus onset, the response will rise from zero to some level; then, as the beam is applied, it will remain steady until the time corresponding to Fig. 6.6 when the response will rise. The fact that the WTA process is not a binary one and that changes occur gradually in an iterative fashion for each level also have impact on the time course of the response. The changes in the unit's response will begin when the WTA is first applied to the configuration of Fig. 6.5 and not only once it completes. Suppose one recorded from one of the units coloured black in layer 3 of Fig. 6.6 throughout (this would correspond to a receptive field that was being attended). One would observe an increment in response late in the response time course here as well. This kind of increment in response well into the attentive process is observed experimentally (for example, Motter, 1993; also, inhibitions were reported); this is the first explanation proposed for it that also can account for inhibition effects of attention. The inhibitory effects are most clearly seen by observing the time course of units coloured white in layers 2 and 3 within the inhibitory beam of Figure 6.6.

The preceding explanation does not by itself account for the serial search observed in experiments of visual search. It does, however, form a part of the explanation. The rest of the explanation includes a method for inhibition of return,

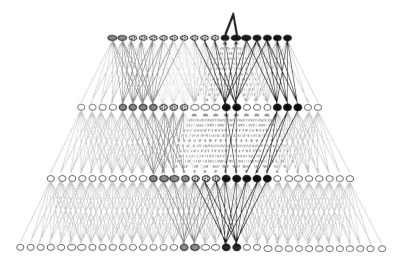

FIGURE 6.5. The second level of attentional inhibition. Several units in layer 3 now receive no input and thus do not provide signals to the output layer.

and for re-deployment of the beam to the next most salient targets in order. The model does in fact accomplish serial search well and these examples can be found elsewhere (Tsotsos et al., 1995). However, there is a body of single-cell recording work which this example does explain. One of those key experiments was that described by Moran and Desimone (1985). What they showed, which was surprising at the time, was that even though an effective stimulus was within a neurone's receptive field, it did not cause the neurone to fire well if the monkey was cued to attend a noneffective stimulus within the same receptive field. A more specific demonstration using the selective tuning model of that exact experimental setup appears in Tsotsos et al. (1995).

From the previous example, it should be clear that the retinotopic distance between the two stimuli in the input layer is important. If the dark gray stimulus were one unit closer to the black, none of its signal would reach the output layer after the application of the attentional beam. On the other hand, if it were one unit farther away, the conflict region should be smaller. Because distance between stimuli is important, if the attended stimulus is near but not in the receptive field studied, the inhibitory effect of attention on the recorded neurone should be large. If it is far, the effect should disappear, and in between the inhibitory effect of attention will gradually decrease with increasing distance. This should be clear from the preceding figures.

Motter (1994a) concluded that the topographic representation of the neural activity in area V4 highlights potential candidates for matching to targets while minimizing the impact of any background items. In other words, the computations that create this representation seem to maximize signal-to-noise ratios for the features which are relevant to the task. Neural activity was attenuated when the stimulus did not match a cue, independent of spatial location, but was about twice as large

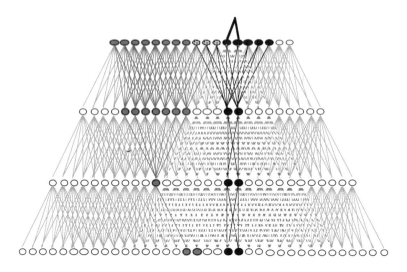

FIGURE 6.6. The third and final level of inhibition due to the attentional beam.

as the attenuated value if the stimulus and cue did match. Motter used color and luminance as features. Interestingly, he found that neural activity was not affected due to the cueing conditions prior to presentation of stimulus arrays. This is consistent with a model that deemphasizes connections that are not of interest. Motter (1994b) goes one step further and concludes that the attentional control system seems to be able to "shut down" the synaptic impact of all but one of many color inputs. This too is consistent with the selective-tuning model and was suggested in Tsotsos (1990) as an important search optimization. Finally, Motter suggests that a sequential combination of a full-field preattentive selection based on features that identify candidate targets, followed by a spatially restrictive focal attentive process which localizes targets, would be an interesting explanation of both his and Moran and Desimone's results; this is exactly the concept initially sketched out in Tsotsos (1990) and embodied in the selective tuning model presented here.

It is important to note that the model is implemented and runs not only in simulation but also controls the attentive behavior of a robotic stereo head. Figure 6.7 shows a hypothetical example of an image of letters, where the system attends to each letter in sequence. This illustrates the method using a two-dimensional example.

The selective tuning model was derived in a first-principles fashion. The major contributor to those principles derives from a series of formal analyses performed within the theory of computational complexity, the most appropriate theoretical foundation to address the question "Why is attention necessary for perception?" The model not only displays performance compatible with experimental observations but also does so in a self-contained manner. In other words, input to the model is a set of real, digitized images and not pre-processed data. The predictive power of the model seems broad:

- An early prediction (Tsotsos, 1990) was that attention seems necessary at any level of processing where a many-to-one mapping of neurones is found. Further, attention occurs in all the areas in concert. The prediction was made at a time when good evidence for attentional modulation was known for area V4 only (Moran and Desimone, 1985). Since then, attentional modulation has been found in many other areas both earlier and later in the visual processing stream, and that it occurs in these areas simultaneously (Kastner et al., 1998). Vanduffel and colleagues (Vanduffel et al., 2000) have shown that attentional modulation appears as early as the LGN. The prediction that attention modulates all cortical and perhaps even subcortical levels of processing has been borne out by recent work from several groups in the 1990's (e.g., Brefzynski and DeYoe, 1999; Ghandhi et al., 1999; Tootell at al., 1999).

- The notions of competition between stimuli and of attentional modulation of this competition were also early components of the model (Tsotsos, 1990) and these too have gained substantial support over the years (Desimone and Duncan, 1995; Kastner et al., 1998; Reynolds et al., 1999).

- The model predicts an inhibitory surround that impairs perception around the focus of attention (Tsotsos, 1990). This, too, has more recently gained support (Caputo and Guerra, 1998; Bahcall and Kowler, 1999; Vanduffel et al., 2000).

- The model further implies that preattentive and attentive visual processing occur in the same neural substrate, which contrasts with the traditional view that these are wholly independent mechanisms. This point of view has also been gaining ground (Joseph et al., 1997; Yeshurun and Carrasco, 1999).

- A final prediction is that attentional guidance and control are integrated into the visual processing hierarchy rather than being centralized in some external brain structure. This implies that the latency of attentional modulations decreases from lower to higher visual areas, and constitutes one of the strongest predictions of the model.

Attention is an important mechanism at any level of processing where one finds a many-to-one convergence of neural inputs and thus potential stimulus interference, a conclusion reached by Tsotsos (1990). This was disputed at first (Desimone, 1990); however, more recent experimental work would appear to be supportive (e.g., Kastner et al., 1998; Vanduffel et al., 2000).

6.6   Conclusions

Marr (1982) had presented the view that a perceptual theory must be defined in three levels: the computational level, the algorithmic level and the implementation level.

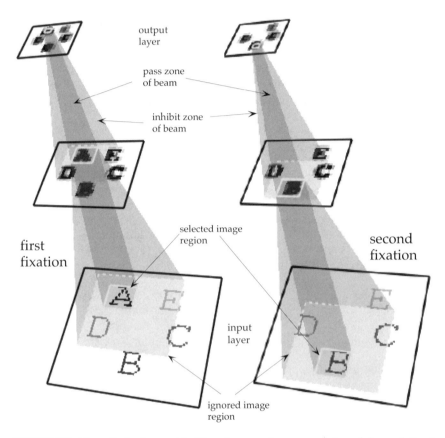

FIGURE 6.7. A hypothetical pyramid of three layers and a representation of shapes (letters) each with a different luminance. In other words, each letter is made up of pixels all of which are a uniform level of brightness and each letter is of a unique brightness. The selective tuning algorithm finds each in luminance order using the strategy described in the text. The inhibitory and pass zones of the inhibitory beam are clearly seen in the first two attentional fixations shown. The computations of each of the two layers above the input layer are based on Gaussian smoothing.

Further, he believed that these levels although related, could be addressed largely independently of one another. What has become very clear since Marr's classic text is that this assertion is false. The computational complexity of a proposed solution is not simply an implementation detail to worry about when coding a solution. If a proposal at the computational level is not realizable due to inherent intractability, then it serves no useful purpose. Because Marr stressed generality of solution, and since it is the case that many basic perceptual problems have already been identified as intractable, the possibility of producing unrealizable solutions following Marr's guidance is great.

This chapter presented a brief overview of the theory of computational complexity with the goals of motivating its importance, explaining its applicability to perception in general, and arguing why this discrete theory is applicable to the study of biological systems. To strengthen the point, a number of very basic sub-problems of perception were used as examples of provably intractable problems. The ultimate claim that results is that neither the brain nor any other implementation can be solving perception in its general form in an optimal manner. A search for optimal solutions to the general problem of perception is thus arguably futile. Guidelines were presented for how to deal with intractable problems and an application of those guidelines to vision in the form of a complexity level of analysis was overviewed. The utility of such an analysis is demonstrated by the development of a model of visual attention that results from that application. A new example of the performance of that model is given that shows how attention might function in the face of conflicting stimuli from a simulated single-cell recording perspective. Several more examples using the same paradigm as well using the implemented model with real images obtained from a robot head can be found in Tsotsos et al. (1995).

What remains? A great deal! Among other problems, complexity-level analysis is expected to provide insights into the following questions:

1. How much information can be extracted from a given attentional fixation?

2. How are successive fixations integrated?

3. How many successive fixations may be processed simultaneously?

4. How does the eye-movement system interact with the covert-attentional system?

5. how is task information represented, and how much of it can be used to tune the visual-processing pyramid?

6. How large is the visual-processing pyramid in terms of numbers of layers, sizes of layers, and number of units computing different visual qualities at each position? Note that all of these open problems depend strongly on the amount of computation that can be performed in a given amount of time, or how much memory is required to store information; thus, complexity is an appropriate tool for their analysis.

Complexity analysis is by no means the most useful tool in the repertoire of the visual scientist. It is, however, a long neglected one, and a critical tool that can predict the realizability and performance of a given perceptual theory with respect to its neural or silicon implementation.

References

Bahcall, D., and Kowler, E., (1999). Attentional Interference at Small Spatial Separations. *Vis. Res.* 39:71-86.

Brefczynski, J. A., and DeYoe, E. A., (1999). A physiological correlate of the 'spotlight' of visual attention. *Nat. Neurosci.* 2:370-4

Caputo, G., and Guerra, S., (1998). Attentional Selection by Distractor Suppression. *Vis. Res.* 38:669-689.

Chelazzi, L., Miller, E., Duncan, J., and Desimone, R., (1993). A neural basis for visual search in inferior temporal cortex. *Nature*, 363:345-347.

Cook, S. (1971). The complexity of theorem-proving procedures. Proceedings of the 3rd Annual ACM Symposium on the Theory of Computing, p. 151-158, New York, NY: ACM Press.

Cooper, M., (1992). *Visual Occlusion and the Interpretation of Ambiguous Pictures*. Chicester, England: Ellis Horwood.

Davis, M.. (1958). *Computability and Unsolvability*. New York, NY: McGraw-Hill.

Davis, M., (1965). *The Undecidable*. New York, NY: Hewlett Raven Press.

Dendris, N. D., Kalafatis, I. A., and Kirousis, L. M., (1994). An efficient parallel algorithm for geometrically characterising drawings of a class of 3-D objects. *J. Math. Imaging Vis.*, 4:375-387.

Desimone, R., (1990). Complexity at the Neuronal Level. *Behav. Brain Sci.*, 13:446.

Desimone, R., and Duncan, J., (1995). Neural Mechanisms of Selective Attention. *Ann. Rev. Neurosci.*, 18:193-222.

Dowling, J. (1987). *The Retina: An Approachable Part of the Brain*. Cambridge, MA: Harvard University Press.

Gandhi, S. P., Heeger, D. J., and Boynton, G. M., (1999). Spatial attention affects brain activity in human primary visual cortex. *Proc. Natl. Acad. Sci. USA*, 96:3314-3319.

Garey, M., and Johnson, D. (1979). *Computers and Intractability: A Guide to the Theory of NP-completeness*. San Francisco: Freeman.

von Helmholtz, H., (1963) in J.P.C.S. Southall (Ed.), *Handbook of Physiological Optics*. New York, NY: Dover, (originally published 1867).

Joseph, J., Chun, M., and Nakayama, K., (1997). Attentional Requirements in a 'Preattentive' Feature Search Task, *Nature* 387:805-807.

Judd, J. S., (1990). *Neural Network Design and the Complexity of Learning.* Cambridge, MA: MIT Press.

Kasif, S., (1990). On the parallel complexity of discrete relaxation in constraint satisfaction networks. *Artif. Intell.*, 45:275-286.

Kastner, S., De Weerd, P., Desimone, R., and Ungerleider, L., (1998). Mechanisms of Directed Attention in the Human Extrastriate Cortex as Revealed by Functional MRI. *Science* 282:108-111.

Kirousis, L. M. (1990) Effectively labeling planar projections of polyhedra, *IEEE Trans. Pattern Anal. Mach. Intell.* 12:123-130.

Kirousis, L. M. (1993). Fast parallel constraint satisfaction. *Artif, Intell.*, 64:147-160.

Kirousis, L.M., and Papadimitriou, C. (1988). The complexity of recognizing polyhedral scenes. *J. Comp. Sys. Sci.*, 37:14-38.

Marr, D. (1982). *Vision: A Computational Investigation Into the Human Representation and Processing of Visual Information.* San Francisco, CA: Freeman.

Maunsell, J., (1995). The brain's visual world: Representation of visual targets in cerebral cortex. *Science*, 270:764-769.

Moran, J., Desimone, R. (1985). Selective attention gates visual processing in the extrastriate cortex. *Science*, 229:782-784.

Motter, B., (1993). Focal attention produces spatially selective processing in visual cortical areas V1, V2 and V4 in the presence of competing stimuli. *J. Neurophysiol..* 70:909-919.

Motter, B., (1994a). Neural correlates of attentive selection of color or luminance in extrastriate area V4. *J. Neurosci.*, 14:2178-2189.

Motter, B., (1994b). Neural correlates of feature selective memory and pop-out in extrastriate area V4. *J. Neurosci.*, 14:2190-2199.

Parodi, P., Lanciwicki, R., Vijh, A., and Tsotsos, J. K. (1998). Emperically-derived estimates of the complexity of labeling line drawings of polyhedral scenes. *Artif. Intel.*, 105:47-75.

Rabbitt, P. (1978). Sorting, categorization and visual search. In E. Carterette, and M. Freidman (Eds.), *Handbook of Perception: Perceptual Processing, (Vol. IX)*, pp. 85-136. New York, NY: Academic Press.

Reynolds, J., Chelazzi, L., and Desimone, R., (1999). Competitive Mechanisms Subserve Attention in Macaque Areas V2 and V4, *J. Neurosci.*, 19:1736-1753

Schall, J., and Hanes, D. (1993). Neural basis of saccade target selection in frontal eye field during visual search. *Nature* 366:467-469.

Stockmeyer, L., and Chandra, A. (1979). Intrinsically difficult problems. *Scientific American Trends in Computing, (Vol. 1)*, pp. 88-97. New York, NY: Scientific American Inc.

Treisman, A. (1988). Features and objects. The fourtennth Bartlett memorial lecture. *Q. J. Exp. Psych.*, 4aA-2:201-237.

Tsotsos, J. K. (1989). The Complexity of Perceptual Search Tasks. Proc. Eleventh International Joint Conference on Artificial Intelligence, Detroit, Michigan, pp. 1571-1577.

Tsotsos, J. K. (1990). A Complexity Level Analysis of Vision. *Behav. and Brain Sci.*, 13:423-455.

Tsotsos, J. K. (1992). On the relative complexity of active vs passive visual search. *Int. J. Comp. Vis.*, 7:127-141.

Tsotsos, J. K. (1995). Behaviorist intelligence and the scaling problem. *Artif. Intell.*, 75:135-160.

Tsotsos, J. K., Culhane, S., Wai, W., Lai, Y., Davis, N., and Nuflo, F. (1995). Modeling visual attention via selective tuning. *Artif. Intell.*, 78:507-545.

Turing, A. (1937). On computable numbers with an application to the Entscheidungs problem. Proc. London Mathematical Society, Vol. 22, 230-265.

Yashuhara, A., (1971). *Recursive Function Theory and Logic*, New York, NY: Academic Press.

Ye, Y., and Tsotsos, J. K., (1996) Sensor planning in 3D object search: its formulation and complexity. Proc. Int. Symposium on Artificial Intelligence and Mathematics, Fort Lauderdale.

Vanduffel, W., Tootell, R., and Orban, G. (2000). "Attention-dependent suppression of metabolic activity in the early stages of the macaque visual system. *Cerebral Cortex*, 10:109-126.

Van Essen, D., Anderson, C., and Felleman, D. (1992). Information processing in the primate visual system: An integrated systems perspective. *Science*, 255:419-422.

Yeshurun, Y., and Carrasco, M., (1999). Spatial attention improves performance in spatial resolution tasks. *Vis. Res.*, 39:293-306

7

Motion-Disparity Interaction and the Scaling of Stereoscopic Disparity

Michael S. Landy
Eli Brenner

Contemporary studies of visual perception often view the observer's problem as one of *estimation*. In other words, the observer seeks to estimate various aspects of the scene, such as the size and shape of the objects in that scene. The information available to inform this estimation is viewed as consisting of one or more visual cues. Each cue may be used on its own to estimate some aspects of scene geometry. Most cues require additional information (visual or otherwise) for the information from that cue to be fully interpreted.

This chapter concentrates on the visual cue to depth of binocular disparity. The cue is the disparity in location of features in the two eyes. Horizontal disparity, the difference in horizontal position in each of the two eyes' views of objects in a scene, is a powerful visual cue to the three-dimensional structure of the world. However, horizontal disparities cannot be fully interpreted without some knowledge of the viewing geometry (the locations, gaze directions, and torsional states of the two eyes). For central gaze (fixating straight ahead), their interpretation requires an estimate of the distance to the fixation point (the *fixation distance*).

Most cues to depth require estimates of the viewing geometry to produce metric estimates of depth in a scene. In a model of depth-cue combination (Landy et al., 1995), this process is described as *cue promotion*. Cues are promoted by the insertion of the values of the unknown viewing geometry and resolution of depth ambiguities. Without promoting the cues, their raw data (e.g., disparities and velocities) are in different units so that simple cue-combination strategies, such as averaging the depth estimates made using each cue, are impossible. When the missing parameters are the eye positions (vergence, gaze directions, and torsions), the promotion process is referred to as *depth scaling*. In particular, in central gaze, the raw sensory data for the cue (velocities, disparities, etc.) are scaled by (that is, multiplied by, or multiplied by the square of) an estimate of the fixation distance. To the extent that this scaling is done accurately, the result is *depth constancy*: perceived depth that is independent of changes in viewing conditions. In this hapter we will limit our discussion of cue promotion to the issue of scaling by the fixation distance.

We review a number of ways in which depth scaling may be accomplished. We then summarize a series of studies of one such strategy involving combination of horizontal disparity with another depth cue, relative motion, to improve the scaling of horizontal disparity. The addition of motion to a stereo display only improves the interpretation of binocular disparities under very circumscribed conditions: it improves shape perception of the moving object (but no other attributes of the percept of that object) and improves shape perception of nearby objects only if they are very similar (in size, shape, and distance) to the moving object. Thus, it appears unlikely that the interaction between the motion and disparity cues leads to an improvement in the estimate of the fixation distance used to scale disparities and other aspects of the 3-D percept.

7.1   Cue Combination in Depth Perception

Researchers in depth perception describe the information that helps observers estimate depth in terms of individual depth cues such as motion, binocular disparities, vergence, and so on. Each cue can potentially provide independent information concerning the layout of a scene. A thorough listing of such cues numbers well over a dozen (Kaufman, 1974).

Although these depth cues are interesting in their own right, and huge numbers of studies have been done on many of the cues, it is also interesting to understand how observers behave when confronted with multiple cues to depth for the same visual judgment. As we will see, there has been increasing attention given to this problem of cue combination over the years. This problem of combining information from multiple sources has also arisen in computer vision, where it is referred to as the depth fusion or sensor-fusion problem (Aloimonos and Shulman, 1989; Clark and Yuille, 1990). Clark and Yuille (1990) describe different ways in which multiple cues can be combined, ranging from weak fusion, where each cue is used to derive an estimate of depth and then the cues are linearly combined, to strong fusion, which allows for arbitrary nonlinear interactions between the cues.

The problem of combining cues is complicated. Different cues provide different types of information about scene layout. At one extreme is the cue of occlusion, which only gives ordinal depth information at occlusion boundaries. At the other extreme is the cue of vergence, which, at least theoretically, when combined with the other gaze parameters (e.g., version, or gaze azimuth) provides an absolute indication of the distance to the fixated object. In between these extremes lie most of the other cues, which often need to be scaled by an estimate of the fixation distance and/or other viewing parameters to provide metric depth values, and some of which are subject to depth reversals.

The modified weak fusion (MWF) model of depth cue combination Landy et al. (1995) was introduced to describe how a weak-fusion rule (depth-cue averaging) could work for perceived depth. It is based on four principles for depth-cue combination: (1) depth cues are linearly combined using a weighted average of the individual estimates derived from each cue; (2) depth-cue weights are based on estimates of the cue reliabilities, so that more reliable cues are given greater weight; (3) depth-cue weights are also based on cue consistency and discrepant depth estimates are downweighted, resulting in a robust overall depth estimate; and (4) depth-cues are not averaged until individual cues are promoted to be on the same scale.

The MWF model has been described so far in normative terms (i.e., what characteristics any cue combination rule *should* rationally have. However, some empirical evidence has been gathered for it as a model of human cue combination. A number of researchers have found depth-cue combination to be linear (e.g., Braustein, 1968; Bruno and Cutting, 1988; Cutting and Millard, 1984; Dosher, Sperling and Wurst, 1986; Johnston, Cumming and Parker, 1993; Young, Landy and Maloney, 1993), although others have disputed whether this is really so, while still others even wonder whether 3D shape is determined on the basis of perceived depth at all, rather than on the basis of surface-centered measures such as curvature (Bülthoff and Mallot, 1988; Curran and Johnston, 1994). We have found that cue weights depend on cue reliability (Young et al., 1993). We also have found some evidence for robust cue combination (Li, Maloney and Landy, 1997).

This chapter centers on the issue of cue interaction for cue promotion. Depth cues provide different kinds of information. Object motion as a cue to depth (the *kinetic depth effect*, Wallach and O'Connell, 1953) only provides relative depth information. The actual metric depth of the object must be scaled by an estimate of the fixation distance. Kinetic depth stimuli also undergo depth reversals. Binocular disparities also must be scaled, but in this case depth is approximately proportional to the square of the fixation distance. Thus, the raw data (velocities and disparities) are effectively in different units and until they are scaled by an estimate of the fixation distance, averaging them is a meaningless operation (i.e., the results will depend on the units of measurement chosen). Thus, combinations of these cues rely first on scaling the individual cues using an estimate of the fixation distance.

Some areas of depth vision and cue combination are not easily handled by the preceding model. For example, a number of cues result in stimuli with depth ambiguities. There are regularities in how observers interpret such stimuli, which can be modeled as the result of a priori biases that the observer brings to the depth-interpretation problem. A Bayesian approach to these ambiguities can be brought to bear to estimate the strength and other parameters of such biases (Mamassian and Landy, 1998) as well as how cues with different biases interact to disambiguate a multicue stimulus (Mamassian, Landy and Maloney, in press).

7.2   Depth Scaling

In this section, we review a number of recent studies of depth scaling, with a particular emphasis on the scaling of horizontal disparities and the sources of information used to estimate the fixation distance.

7.2.1   Failures of depth constancy with stereo

A number of researchers have examined whether there is depth constancy for binocular disparities. Wallach and Zuckerman (1963) found good but incomplete constancy with changes of distance when they provided only accommodation and vergence cues to distance consonant with the change in optical distance to the display. Ono and Comerford (1977) describe a number of early studies of depth constancy and review a number of theories. The main result of several studies of the perception of depth from disparity at multiple distances is generally summarized as partial constancy due to a misestimate of the distance (Foley, 1980; Johnston, 1991). Johnston (1991) introduced the *apparently circular cylinder* (ACC) task for examining this issue. In this task, subjects are presented with computer renderings of cylinders with elliptical cross sections, and it is experimentally determined how much depth is required for the cylinders to appear to the subjects to have a circular cross section. For moderately sized, random-dot stereograms, shapes appear distorted in a manner consistent with the hypothesis that observers misjudge the distance, exaggerating it (and perceived depth) when it is substantially less than 1 m, and underestimating it when it is large. Collett, Schwartz, and Sobel (1991) found similar results, with the addition that the size of an object had an impact on scaling; smaller objects were treated as if they were located farther away (and hence were scaled so as to be larger physically and have greater disparity-defined depth).

7.2.2   Distance scaling of size, shape, and depth

Depth is not the only perceptual variable that depends on the viewing geometry. A fixed retinal object should increase in both perceived depth and perceived size (height and width) with an increase in estimated distance. The size should be proportional to the distance and, to first order, the depth proportional to the square of the distance. Hence, aspects of perceived shape (e.g., depth/width, which is the relevant variable in the ACC task) also depend on the viewing distance.

Logically, one might expect the visual system to estimate the viewing geometry as the 6 degrees of freedom of eye position, 3 for each eye, although the binocular extension of Listing's Law implies that there are only 3 degrees of freedom for binocular eye movements toward binocularly fixated targets (see, e.g., van Rijn and van den Berg, 1993), and then use this estimate to scale all items that depend on it (depth, size, shape, etc.). An understanding of scaling necessarily involves several questions. What cues are available to estimate the viewing geometry? How do observers use, weight and combine these cues to estimate the fixation distance and other aspects of viewing geometry? Does this result in a single estimate of viewing geometry that is then used to scale all measurements that require such scaling?

If we restrict ourselves to central gaze, then the only viewing parameter needed is the fixation distance. We will refer to the estimate of the fixation distance used to scale disparities as the *scaling distance*. Several cues to the fixation distance could conceivably be used. Johnston (1991) found partial constancy in a reduced-cue situation in which vergence and accommodation were the primary cues to distance. There has been some controversy over whether the pattern of vertical disparities between the two eyes' images used for scaling. Helmholtz (1910) was the first to demonstrate the use of vertical disparities using the apparent frontoparallel plane task. Cumming, Johnston and Parker (1991) and Sobel and Collett (1991) concluded that vertical disparities are not used, and Rogers and Bradshaw (1993) and others concluded that they are, with the primary distinction being the size of the display (larger displays yield larger, perhaps more reliable, vertical disparities, Bradshaw, Glennerster, and Rogers, 1996). Another cue involves combining two depth cues that scale differently with the viewing distance (e.g., stereo/motion interaction), a strategy that is the focus of this hapter. Collett, Schwarz, and Sobel (1991) found that relative size affected the scaling distance. Other cues could be used as well, such as the observer's knowledge of the actual distance to the CRTs used in a given experiment or a default value for distance in the absence of other information (*the specific distance tendency* – Gogel and Tietz, 1973).

Given the multiple cues to the fixation distance, its estimation constitutes a cue combination problem to which the principles of MWF might be applied. In our work on stereo/motion interaction, reviewed in Section 7.3.3, we found that the scaling distance was a compromise between that indicated by stereo/motion interaction and that indicated by other cues such as vergence or prior knowledge (Econopouly and Land, 1995). Bradshaw, Glennerster, and Rogers (1996) provide evidence of a weighted combination of vergence and vertical disparity cues to distance, with vertical disparity receiving higher weight with larger displays, which provide larger, more reliable vertical disparities.

The final question is whether these cues, once combined, result in a single value of scaling distance that is then used for all scaling problems that require such an estimate: shape, size, depth, and apparent distance. When one manipulates cues to the fixation distance, there are changes to all of these perceptual attributes, suggesting that there is common distance information used to make all of these perceptual estimates (Rogers and Bradshaw, 1993; van Damme and Brenner, 1997; Brenner and van Damme, 1999).

7.3   Stereomotion Interaction for Depth Scaling

For some time now, we have been examining whether observers combine the depth cues of binocular disparity and object motion to help determine the distance that is then used to scale these cues. The rest of this chapter is concerned with this particular cue interaction. We begin by reviewing why these two cues might be useful in determining the distance. Then, we review the evidence we have gathered as to the circumstances in which it is and is not used.

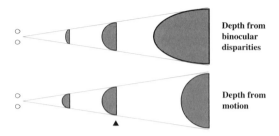

Veridical scaling distance

FIGURE 7.1. The scaling problem. Depth from binocular disparities scales approximately with the square of the distance. Thus, a given set of retinal disparities is consistent with a nearby, squashed ellipsoid, a circular cross section at an intermediate distance, or an ellipsoid stretched in depth at a far distance. Depth from relative motion scales linearly with distance. Thus, a motion display consistent with a circular cross section at a nearby distance is also consistent with a larger, but still circular, cross section at farther distances. An interaction of stereo and motion (Richards, 1985; Johnston, Cumming, and Landy, 1994) might involve choosing the distance for which the two cues are consistent, resulting in motion determining the shape and consistency determining the size and distance.

7.3.1   Why combine stereo and motion?

Depth from motion scales linearly with the distance, whereas depth from disparity scales as the square of the fixation distance. Because horizontal extent also scales linearly with the distance, this means that shape from motion (e.g. depth/width) is independent of distance. On the other hand, the same retinal disparities imply different shapes as a function of the distance (Fig. 7.1). Richards (1985) pointed out that one can combine horizontal disparity information with relative motion to derive an estimate of the distance. To do this, one need only select the distance for which the shape estimates from motion and from stereo are in agreement. This cue-interaction scaling hypothesis, were it the only method used for scaling, would result in motion determining the shape (e.g., squashed, circular or stretched in the ACC task). The interaction between motion and disparity would determine the distance and consequently the size.

7.3.2   Evidence with a single object

A number of studies have involved shape judgments with displays that combine binocular disparity and motion cues. Most did not directly address the issue of stereo/motion combination for scaling. For example, Rogers and Collett (1989) added motion parallax to a stereo display of a corrugated surface. The depth percepts from motion alone and disparity alone were in conflict. Observers perceived a curved motion path in the two-cue displays, which, in their stimulus situation, effectively resolved, or at least minimized, the conflict. Tittle and Braunstein (1990) also combined motion and stereo cues, and found evidence for linear cue combination, with an additional finding of cooperation between the two cues that can be interpreted in terms of motion helping the observers to solve the stereo correspondence problem.

Brenner and van Damme (1999) found that shape judgments improve when object motion is added to a stereo display. In their experiment, subjects adjusted the size and depth of an ellipsoid so that it appeared spherical and the size of a tennis ball they held in one of their hands throughout the experiment. Observers also performed a reach to the apparent distance of the rendered object. Rotating the object improved shape settings, but had little effect on size settings and apparent distance. Thus, the distance-independent shape from motion was used for the shape settings, but the conflict with stereo was not used to improve the distance estimate used to scale size or determine apparent distance. On the other hand, turning the room lights on (to improve cues to distance) improved settings of size, shape, and apparent distance. The results suggest that there are common signals to distance but that the three judgments were handled separately.

The hypothesis that motion and stereo are combined to determine the distance was tested using the ACC task with rotating cylinders (Johnston, Cumming and Landy, 1994). Both stereo and motion were available in some of their displays. They were interested in whether stereo/motion combination, when available, was the sole determinant of the distance estimate. We will show later that stereo/motion may have been used, but only in combination with other estimates of distance.

In their study, observers performed the ACC task for cylinders which either included binocular disparities (versus monocular viewing), structure from motion (rotation back and forth), or both. In the stereo-only condition, settings were biased in a manner that mimicked the previous results of Johnston (1991). The motion-only results were generally accurate, reflecting the lack of distance dependence of structure-from-motion for shape. Settings were also accurate in the condition that included both disparities and motion, consistent with the notion that the two cues are combined in such a way that motion determines the shape (the only perceptual attribute probed by the task), and stereo helps solve for distance and object size. However, in a second experiment, the ACC task was used with stereomotion displays in which the two cues signalled differing amounts of depth (a cue-conflict stimulus). As the depth rendered using disparity was increased, the depth from motion required for an apparently circular cylinder decreased. This implied that binocular disparities did indeed have a weight in the calculation of object shape. It is this apparent contradiction (disparity interpretation is determined completely by motion information, but disparities still have a weight in determining object shape) that led us to do the work described in the next section.

Finally, Tittle, Todd, Perotti, and Norman (1995) found that combinations of binocular disparities and motion did not always result in veridical estimation of shape. They also used the ACC task, but found that the results were different depending on the aspect (the average slant) of the rotating object. They interpreted their results as supporting nonmetric (e.g. affine) representation of shape.

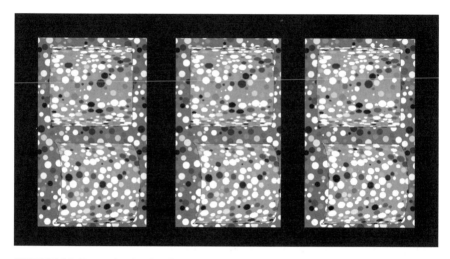

FIGURE 7.2. Example stimulus from a study comparing perceived depth of the top, static cylinder in the context of either a static or rotating bottom cylinder (Econopouly, 1995; Econopouly and Landy, 1995). The left hand image pair is for crossed fusion, and the right hand image pair for diverged fusion.

7.3.3   Two neighboring objects

The results of the study by Johnston, Cumming, and Landy (1994) were puzzling. Their first experiment indicated that the combination of stereo and motion determines the distance used to rescale stereo in a manner intended to make the shape indicated by the two cues consistent. Thus, it appeared that the motion cue determined the perceived shape. The contribution of stereo (in combination with the motion cue), if any, was to determine the distance and size. They did not measure perceived distance and size, and the results of Brenner and van Damme (1999) indicate that stereo/motion combination would not even have had that effect. Because the ACC task only measured perceived shape, then stereo should have had no weight in the results of that task. Direct measurement of cue weights in Johnston et al.'s second experiment revealed a substantial weight for stereo in the combination. How can this be?

One possibility is that motion/stereo combination contributes to scaling but is not the sole source of distance information. We decided to pursue this possibility by adding a second, static object to the rendered scene (Econopouly, 1995; Econopouly and Landy, 1995). The scaling distance used by observers was measured by having them judge the shape of the static, stereo object in the context of a nearby, rotating one. If the interaction between motion and stereo did indeed rescale stereo disparities for these rotating cylinders by helping the observer determine a more accurate estimate of the fixation distance, this improved distance estimate should logically be used to rescale all stereo disparities in the scene. There is, after all, only one fixation distance at a time.

Observers viewed textured, horizontally-oriented cylinders rendered using ray-

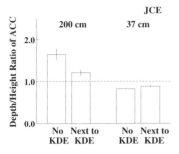
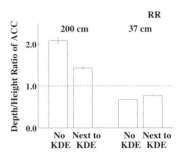
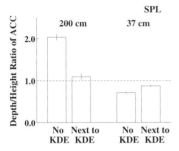
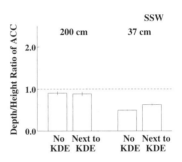
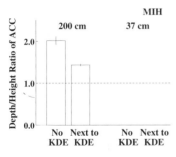

FIGURE 7.3. Apparently circular cylinders (estimated from psychometric functions for forced-choice judgments) for five observers (Econopouly, 1995). For the static cylinders viewed next to a second, static cylinder, the results replicate those of Johnston (1991). In other words, at the 37 cm viewing distance, the depth/height ratio of the ACC was less than 1, implying that observers perceived exaggerated depth (as if they were overestimating the viewing distance when scaling disparity), and at the 200 cm viewing distance the opposite was found. In nearly all cases, the static cylinder was perceived in a more accurate fashion in the context of an adjacent, rotating cylinder (the exception was observer SSW at the 200 cm distance, for which there was no effect of context).

tracing from the optically defined viewpoints of the two eyes (i.e., taking into account the optical path including the stereoscope mirrors). A far (200 cm) and a near (37 cm) distance were used at which, according to the results of Johnston (1991), the distance and hence the depth should be under- and over-estimated, respectively. Two cylinders were shown, one above the other on a static, flat, textured background (Fig. 7.2). The lower cylinder was always rendered with a circular cross section. In one condition, both cylinders were static. In the other condition, the lower cylinder rotated back and forth about a vertical axis. The upper, static cylinder's depth (indicated primarily by binocular disparities, but also by texture and occluding contour) was varied across trials. Observers judged whether the upper cylinder appeared to be stretched or squashed in depth relative to a cylinder with circular cross section (that is, the ACC task). The ACC was estimated as the 50% point on the resulting psychometric function.

The results are clear (Fig. 7.3). When a static cylinder is judged and the adjacent cylinder is also static, binocular disparities are misscaled in a manner that replicates Johnston's findings (1991). When one of the cylinders rotates, not only is it perceived more veridically (Johnston, Cumming, and Landy, 1994), but so is the adjacent, static cylinder. This is completely consistent with the idea that an interaction between motion and disparity in the lower cylinder is used to improve the observer's estimate of the distance, which is in turn used to scale all of the disparities in the scene.

The results of a second experiment (Econopouly, 1995) will allow us to reject several alternative explanations for these context effects. In this experiment, the lower, rotating cylinder was not always rendered as circular in cross section. In Table 7.1, we show the results for conditions in which the rotating cylinder had *consistent cues*. That is, the rotating cylinder was rendered veridically, with binocular disparities and motion both indicating the same amount of depth. The second column shows results when the lower cylinder was static (taken from the previous experiment). They are given in terms of the *effective distance*, which is the distance at which the horizontal disparities of the ACC would be consistent with a rendered circular cylinder. For all three subjects, the effective distances are shorter than the actual viewing distance of 200 cm, which is consistent with the results of Johnston (1991). The third column describes the depth rendered for each cue in the rotating cylinder in units of the cylinder's half-height (so that a circular cylinder corresponds to a value of 1.0). The third column also provides the *equivalent distance*, which is the distance at which the velocities and horizontal disparities are consistent with the same shape as one another (that is, the distance indicated by stereo-motion interaction). For these consistent cue, rotating cylinders, the equivalent distance is the distance for which the stimuli were rendered (i.e., 200 cm). The ACC for the upper, static cylinder was also measured in the context of the rotating cylinder described in the third column whose depth could either be flatter than a circular cylinder (a depth of 0.5), circular (1.0), or exaggerated (1.5 or 2.0). For all three subjects, the upper cylinder ACC setting improved (became more veridical), which can be seen in Table 7.1 as the effective distance becoming closer to the correct value of 200.

TABLE 7.1. ACC results of Econopouly (1995) for static cylinders in the context of rotating cylinders with consistent depth indicated by motion (d_{KDE}) and disparity (d_{BD}) viewed from 200 cm. The values of d_{KDE} and d_{BD} are in units of the cylinder's half-height, so that a value of 1.0 is the depth that would result in a circular cylinder. The second column gives the effective distance for the ACC setting of the upper cylinder when the lower cylinder was static. The third column describes the rendered depth of the lower cylinder. In this table the cues for this cylinder were consistent, so that the equivalent distance was the same as the rendered distance of 200 cm. The fourth column gives the effective distance for the ACC settings of the upper, static cylinder when the lower cylinder was rotating.

Subj.	Effective Distance (No KDE)	Equivalent Distance from BD & KDE		Effective Distance (Next to KDE)
JCE	121 cm	$d_{BD} = 0.5$ $d_{KDE} = 0.5$	200 cm	140 cm
		$d_{BD} = 1.0$ $d_{KDE} = 1.0$	200 cm	145 cm
		$d_{BD} = 1.5$ $d_{KDE} = 1.5$	200 cm	146 cm
		$d_{BD} = 2.0$ $d_{KDE} = 2.0$	200 cm	141 cm
SPL	98 cm	$d_{BD} = 0.5$ $d_{KDE} = 0.5$	200 cm	142 cm
		$d_{BD} = 1.0$ $d_{KDE} = 1.0$	200 cm	148 cm
		$d_{BD} = 1.5$ $d_{KDE} = 1.5$	200 cm	141 cm
		$d_{BD} = 2.0$ $d_{KDE} = 2.0$	200 cm	146 cm
RR	96 cm	$d_{BD} = 0.5$ $d_{KDE} = 0.5$	200 cm	183 cm
		$d_{BD} = 1.0$ $d_{KDE} = 1.0$	200 cm	182 cm
		$d_{BD} = 1.5$ $d_{KDE} = 1.5$	200 cm	172 cm
		$d_{BD} = 2.0$ $d_{KDE} = 2.0$	200 cm	169 cm

There was little or no trend in these ACC settings as a function of the depth of the neighboring, rotating cylinder. This finding argues against depth contrast or assimilation effects. The rotating cylinder can be perceived as flatter or more extended in depth than the upper, static cylinder, and yet its rotation will cause the perceived depth of the upper, static cylinder to increase. In the conditions shown in Table 7.1 it is true, however, that the rotation of the lower cylinder always increases the depth of both the rotating cylinder (Johnston, Cumming, and Landy, 1994) and the upper, static one.

In a second set of conditions (Table 7.2), the lower, rotating cylinder had inconsistent cues. In other words, if interpreted at the optical viewing distance (that indicated by the vergence angle), the depth indicated by motion (and by texture and occluding contour, which always agreed with the motion cue) could be different from that indicated by horizontal disparities. This cue conflict was accomplished by supplying the rendering software with a fallacious value of the inter-pupillary distance (as in Johnston, Cumming, and Landy, 1994), allowing us to exaggerate or diminish the disparities independent of the object motion and texture. You can think of these cue-conflict stimuli as a means of perturbing the equivalent distance (the distance estimate from combining the stereo and motion information). For example, in the second row in the table, depth indicated by motion is exaggerated and depth from disparity is halved (at the rendered distance of 200 cm). This combination of disparities and velocities indicate the same shape if the cylinder is interpreted as located much farther away (an equivalent distance of 554 cm).

The set of rotating cylinders was determined in a control experiment; they were all apparently circular cylinders (this control experiment was a replication of Experiment 2 of Johnston, Cumming, and Landy, 1994). For most subjects and conditions, the effective distance for the static, upper cylinder was a compromise between the effective distance without the context and the equivalent distance from the context (the exceptions are rows 4, 8, and 11 in the table). For subject JCE, the context was able to increase or decrease the effective distance. In other words, the rotation of the lower cylinder could either increase perceived depth, moving the effective distance closer to the veridical value of 200 cm, or decrease depth, making the effective distance less veridical than it already was. This was not the case for the two other subjects, for whom the rotation of the lower cylinder always increased perceived depth of the upper, static cylinder. These results are mostly, but not completely, consistent with the idea that stereomotion interaction results in an estimate of the fixation distance that is then combined (e.g., averaged) with other estimates or defaults before being used to scale disparities. Again, these results argue against any explanation based on the perceived depth of the lower, rotating cylinder affecting perceived depth of the upper cylinder (e.g., contrast or assimilation effects), as a wide range of context effects was achieved using rotating cylinders that were all perceived to have the same circular shape.

Finally, the effect of the rotating cylinder is not due to the motion of the adjacent object per se leading to the rescaling. An adjacent monocular rotating cylinder does not lead to any measurable rescaling.

These results help solve the conundrum of the previous section. Rotating one

TABLE 7.2. ACC results of Econopouly (1995) for static cylinders in the context of rotating cylinders for which depth indicated by motion (d_{KDE}) and disparity (d_{BD}) were inconsistent, viewed from 200 cm. d_{KDE} and d_{BD} are in unit's of the cylinder's half-height; $d_{KDE} = d_{BD} = 1.0$ corresponds to a circular cylinder. The particular values of d_{KDE} and d_{BD} were chosen from a control experiment so that all the rotating cylinders would appear to be circular. Thus, it was hoped that any effect of the context on the upper, static cylinder's appearance would be due to stereo-motion interaction and not simply to the appearance of the lower cylinder (which didn't change across conditions).

Subj.	Effective Distance (No KDE)	Equivalent Distance from BD & KDE		Effective Distance (Next to KDE)
JCE	121 cm	$d_{BD} = 0.0$ $d_{KDE} = 1.4$	∞ cm	207 cm
		$d_{BD} = 0.5$ $d_{KDE} = 1.3$	554 cm	171 cm
		$d_{BD} = 1.0$ $d_{KDE} = 1.1$	221 cm	150 cm
		$d_{BD} = 1.5$ $d_{KDE} = 0.9$	116 cm	128 cm
		$d_{BD} = 2.0$ $d_{KDE} = 0.8$	78 cm	100 cm
SPL	98 cm	$d_{BD} = 0.5$ $d_{KDE} = 1.2$	474 cm	168 cm
		$d_{BD} = 1.0$ $d_{KDE} = 0.8$	164 cm	146 cm
		$d_{BD} = 1.5$ $d_{KDE} = 0.4$	54 cm	134 cm
RR	96 cm	$d_{BD} = 0.5$ $d_{KDE} = 1.2$	488 cm	213 cm
		$d_{BD} = 1.0$ $d_{KDE} = 0.9$	182 cm	175 cm
		$d_{BD} = 1.5$ $d_{KDE} = 0.5$	111 cm	154 cm

cylinder causes the adjacent cylinder to be rescaled so that it appears *more* veridical, but not *completely* veridical. This is true for all of the results in Table 7.1, and 8 of the 11 results in Table 7.2. Thus, the shape derived from disparity is different from that derived from motion, and so any cue-combination rule such as the weighted average we usually find (Landy et al., 1995) could still show a non-zero weight for the disparity cue. This was not seen in the previous data (Johnston, Cumming, and Landy, 1994) because the difference between the stereo-derived shape and the motion-derived shape was less than the variability of the measurements.

7.3.4   *Two objects and alternative computations*

Thus far, the evidence suggests that stereomotion combination improves the scaling distance. Are there alternative explanations? Suppose that subjects do none of the things we have attributed to them: they do not use stereomotion interaction to estimate the viewing distance, and they do not use this improved distance estimate on either the rotating object or the nearby, static object. Is it possible to account for our results some other way?

Suppose that subjects do not know how to scale disparities. Instead, suppose that their behavior in the various ACC and depth-comparison tasks we have described has been carried soley out by tricks specific to the particular experimental conditions. For a single object (Johnston, Cumming, and Landy, 1994), the rotation might result in a stimulus for which subjects sense cue conflict. They might then set the disparities based on remembered disparities for rotating cylinders they have previously experienced at the given distance. Thus, this disparity correction might *not* result in a new distance estimate and hence should not affect other judgments of this or other objects.

For the case of two objects (Econopouly, 1995), consider a static, stereo cylinder next to a circular, rotating cylinder. In these experiments, the two cylinders lay on a flat, fronto-parallel background, making it abundantly clear that they were located the same distance from the observer. For the results in Fig. 7.3, the rotating cylinder is perceived as approximately circular, therefore the static cylinder, to appear circular, should logically have the same amount of disparity. Thus, observers could have performed the task by, effectively, copying the disparities of the rotating cylinder. The two objects in that study were located at the same distance and had the same width, so that this trick was particularly easy to use (not that subjects were aware of using it).

The required trick is, in fact, slightly more complicated. In Table 7.1, there are conditions in which the rotating cylinder does not have a circular cross section. Thus, we must further posit that an observer viewing, say, a rotating cylinder with twice the depth of a circular cylinder, will halve its disparity before copying it (or comparing it) to the neighboring, static cylinder. For the rotating cylinders containing inconsistent motion and stereo cues (Table 7.2), the rotating cylinder stereo/motion combinations were chosen to result in apparently circular cylinders. The strategy we have described implies that the disparities should have been copied to the static cylinder unmodified, resulting in an effective distance (for the static

FIGURE 7.4. Example stimulus from Brenner and Landy (1999). The upper image pair is for diverged fusion and the lower one for crossed fusion.

cylinder) equal to the equivalent distance (for the rotating one). This was not found, but the effect of changes in the equivalent distance of the rotating cylinder, although too small, *was* in the appropriate direction.

7.3.5 Two objects at unequal distances

More convincing evidence was needed that observers combined stereo and motion to better estimate the fixation distance. We approached this problem (Brenner and Landy, 1999) by elaborating the experimental paradigm first used by Brenner and van Damme (1999). Observers viewed two textured, stereo ellipsoids, side by side, in an otherwise dark room (Fig. 7.4). The rendered distance to the left-hand ellipsoid was varied over a wide range. Subjects were required to make five adjustments to the rendered objects. First, the left-hand ellipsoid was adjusted to appear equal in size and depth to a tennis ball (i.e., 3.3 cm in radius). Next, the distance to the right-hand ellipsoid was adjusted either to be half that of the fixed, left-hand one, or equal to it (in different blocks of trials). Finally, the size and depth of the right-hand ellipsoid were also set to appear equal that of a tennis ball or, in one condition, to double its size. In some trials, the left-hand ellipsoid rotated back and forth about a horizontal axis; in others, it was stationary. The right-hand ellipsoid was always stationary. If stereo/motion combination resulted in an improved distance estimate, then rotation of the left-hand ellipsoid should have improved the accuracy of all of the settings of *both* ellipsoids.

The results gave little support for stereomotion combination helping observers estimate distance (Fig. 7.5). Individual panels of the figure show one observer's individual settings of width and depth for different conditions. A correct setting for the rendered ellipsoids would have been to adjust all stimuli to lie at the point width = depth = 3.3 cm (except for the double-sized condition). A correct setting of object shape (to be spherical) with the size set incorrectly would have resulted in points along the dashed, diagonal lines. For the static condition (Fig. 7.5a), the settings lie closer to the curve. The points along this curve correspond to disparity and retinal-size settings consistent with a sphere of a fixed size (estimated from the data, larger than the correct value of 3.3 cm for all subjects) located at a distance different than that which was rendered. These results are generally consistent with

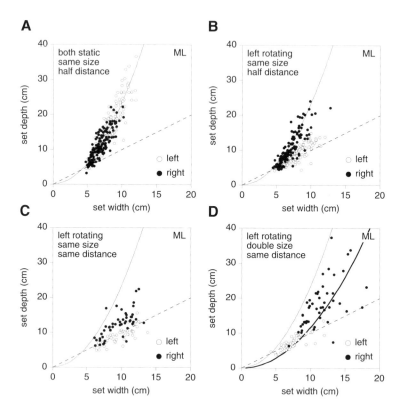

FIGURE 7.5. Results of Brenner and Landy (1999). In each panel, symbols indicate individual settings of half-width and half-depth of the left-hand (open symbols) and right-hand (filled symbols) ellipsoids. Veridical settings would lie at (3.3, 3.3). The dashed line is the locus of true spheres at the rendered distance. The solid curves are the width and depth (at the rendered distance) for which the retinal size and disparity are identical to those of a sphere of a fixed size (not 3.3 cm, but instead for a size determined separately for each subject to fit one condition's data) for a range of distances. It is a curve because the width follows a linear law with distance, and the depth from disparity follows a square law. (a) Both spheres static and the right-hand one set to half the distance of the left-hand one. Settings for both cluster around the curve, consistent with misestimation of the distance. (b) The left-hand sphere is now rotating. Its settings now cluster around the dashed line, and hence are more spherical for the rendered distance. Its size (width) is set no more accurately than before, and settings for the right-hand ellipsoid are not improved. (c) Settings for the case of equal distance. Now both the rotating and static ellipsoids are set to be spherical as in Econopouly (1995). (d) The ellipsoids are set at equal distances, with the right-hand, static one set to be twice the size of a tennis ball. The extra, bold curve corresponds to a double-sized ball at various distances. The scatter is large but appears to follow the bold curve.

Johnston (1991). When the left-hand ellipsoid rotates, its settings now cluster about the diagonal line (panel b, open symbols), indicating a setting that is more spherical, consistent with the ACC results of Johnston, Cumming, and Landy (1994). This is also true for settings of the right-hand, static ellipsoid when it is next to a rotating ellipsoid located at the same distance and set to the same size (panel c, solid symbols), consistent with the results of Econopouly (1995).

But, beyond these replications of previous results, all other aspects of the data in Fig. 7.5 are inconsistent with the idea that the rotation of the left-hand, stereo ellipsoid improves observers' estimate of the distance. There is considerable scatter in the size (width) settings of the left-hand, static ellipsoids (Fig. 7.5a, open symbols). These size settings become no less variable or biased when the ellipsoid rotates (panel b, open symbols), which is consistent with the results of Brenner and van Damme (1999). The shape settings of the right-hand ellipsoid do not become more spherical in the context of a rotating ellipsoid when located at half the distance (panel b, filled symbols). These shape settings also improve very little when the distances are equal but the size of the right-hand ellipsoid is set to be double (panel d, filled symbols). For the condition in which subjects halved the distance, the distance settings were unreliable, biased, and unimproved by the rotation of the left-hand ellipsoid (not shown). The set distance was also inconsistent with the set width, which was surprising, as Brenner and van Damme (1999) found that subjects can copy, double, or halve distances across changes in version (gaze azimuth) when only vergence is available, even though they can not judge absolute distance well at all. Finally, subjects were poor at the size settings across the halved distance, even though all they needed to do was to set the retinal size of the right-hand ellipsoid to be double that of the left-hand one.

To summarize, although we replicated the previous findings of improved shape settings for rotating, stereo objects and for static, stereo objects located next to them, these effects are very restricted. The improvement of shape settings for the neighboring, static object only occurred if it was located at the same distance and had the same size as the rotating object. Also, improvements in shape settings were not associated with improvements in set size or set distance. Thus, there is very little evidence for an underlying, improved estimate of the distance used to scale the various scene attributes that logically require distance scaling.

We remained puzzled by this since the trick we suggested to explain E-conopouly's results does not fully explain them, and subjects are certainly not aware of using such a trick. We ran another experiment (Landy and Brenner, 1999) to try to determine the limits of the effect of context: How similar in size and distance must the two objects have to be for the rotation of one to result in improved shape settings for the other? The methods and stimuli were identical, in most respects, to the previous experiment (Brenner and Landy, 1999). Again, two ellipsoids were rendered side by side. The right-hand ellipsoid was static; the left-hand ellipsoid was static in half the trials and rotating back and forth about a horizontal axis in the other half of the trials. The distance of each was chosen from three possible values (225, 300 or 400 cm), as was the size of each (12.6, 16.8 or 22.4 cm diam). Observers adjusted the depth of each ellipsoid so that it appeared

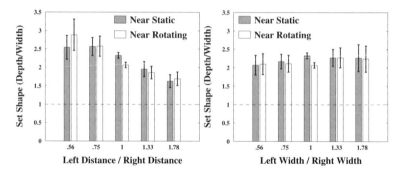

FIGURE 7.6. Shape settings from the study of Landy and Brenner (1999). A value of 1.0 indicates a veridical, spherical setting. Error bars indicate plus or minus two standard errors. Settings for the static, right-hand ellipsoid only became more veridical in the context of a rotating, left-hand ellipsoid when the two ellipsoids were located at the same distance (left panel) and had the same size (right panel).

spherical (thus, there were two adjustments to make per trial). Six observers ran 368 trials each. In about half the trials, the two ellipsoids were the same size and at the same distance. In the other half, the size and distance were chosen at random. As in the previous experiment (see Fig. 7.5, Brenner and Landy, 1999), subjects almost always set the depth too large. Rotating the left-hand ellipsoid reduced this error for the rotating ellipsoid. Figure 7.6 summarizes the shape settings (set depth divided by rendered width) for the other, right-hand ellipsoid. A veridical setting results in a set shape value of 1.0. Again, there was no significant improvement of the shape setting of the static, right-hand ellipsoid in the context of a rotating, stereo ellipsoid unless the two ellipsoids were located at the same distance (left panel; left/right = 1) and had the same size (right panel; left/right = 1). Note that the smaller standard errors in these conditions are due to the large number of trials in which distance and width are identical for the two ellipsoids.

The question remains: When, if ever, does stereomotion combination result in improved scaling, and what object attributes' scaling is improved by it? It is possible that all of the effects described here stem from associations and heuristics as described in Section 7.3.4. It is also possible that the scaling distance is improved, but this improved estimate is only applied under very limited circumstances.

In all of the experiments described here, eye movements were unconstrained. Typically, observers would fixate one object and then the other, changing vergence if they lay at different distances. Perhaps the stereo-motion interaction only applies while the rotating object is actively fixated, or only across iso-vergent saccades from that rotating object. When one judges a second, static object lying at a different distance (Brenner and Landy, 1999), then the improved scaling distance from the rotating object must be combined with information about the required vergence change to fixate the static object to estimate the latter's distance. In Brenner and Landy (1999), the distance was halved, and both objects were never seen fused at the same time. In the final study (Landy and Brenner, 1999), the differences in distance were substantially smaller and the objects were often seen fused simultaneously. Nevertheless, observers changed fixation from object to object to make their settings. In fact, observers find it nearly impossible to do these tasks when asked to maintain fixation on a spot lying off of the object to be judged.

7.4   Summary

It would be logical for observers to estimate scene attributes such as object size, depth, shape, and distance using all the information available to them. For most of these attributes, as estimated by many available cues, complete estimation requires the observer to promote the measurements (relative velocities, shading gradient, binocular disparities, etc.) by parameters related to the viewing geometry. When stereo is the only cue to depth available, shape is often misestimated as if the wrong value of the viewing distance were being used. Shape estimates improve when an object rotates, or when an object is next to an identical, rotating object. But it seems that these are the only estimates of scene attributes that are improved by the presence of a rotating, stereo object. Although the information is available to refine an observer's estimate of the distance using the concurrent stereo and motion information, observers either do not use it at all, or use it only to improve shape estimates under very restricted circumstances. Viewing distance estimates may well be computed using multiple cues (e.g., Bradshaw, Glennerster, and Rogers, 1996), but our evidence suggests that the combination of stereo and motion rarely, if ever, contributes to such estimates.

Acknowledgments

This work was supported by NIH grant EY08266 and the AFOSR. We thank Larry Maloney and Marty Banks for comments on the manuscript.

References

Aloimonos, J., and Shulman, D. (1989). *Integration of Visual Modules: An Extension of the Marr paradigm.* New York: Academic Press.

Bradshaw, M. F., Glennerster, A., and Rogers, B. J. (1996). The effect of display size on disparity scaling from differential perspective and vergence cues. *Vis. Res.*, 36:1255–1264.

Braunstein, M. L. (1968). Motion and texture as sources of slant information. *J. Exp. Psych.*, 78:247–253.

Brenner, E., and van Damme, W. J. M. (1998). Judging distance from ocular convergence. *Vis. Res.*, 38:493–498.

Brenner, E., and van Damme, W. J. M. (1999). Perceived distance, shape and size. *Vis. Res.*, 39:975–986.

Brenner, E., and Landy, M. S. (1999). Interaction between the perceived shape of two objects. *Vis. Res.*, 39:3834–3848.

Bruno, N., and Cutting, J. E. (1988). Minimodularity and the perception of layout. *J. Exp. Psych. Gen.*, 117:161–170.

Bülthoff, H. H., and Mallot, H. A. (1988). Integration of depth modules: stereo and shading. *J. Opt. Soc. Am. A*, 5:1749–1758.

Clark, J., and Yuille, A. (1990). *Data Fusion for Sensory Information Processing Systems*. Boston, MA: Kluwer.

Collett, T. C., Schwarz, U., and Sobel, E. C. (1991). The interaction of oculomotor cues and stimulus size in stereoscopic depth constancy. *Perception*, 20:733–754.

Cumming, B. G., Johnston, E. B., and Parker, A. J. (1991). Vertical disparities and perception of three-dimensional shape. *Nature*, 349:411–413.

Curran, W., and Johnston, A. (1994). Integration of shading and texture cues: Testing the linear model. *Vis. Res.*, 34:1863–1874.

Cutting, J. E., and Millard, R. T. (1984). Three gradients and the perception of flat and curved surfaces. *J. Exp. Psych. Gen.*, 113:198–216.

van Damme, W., and Brenner, E. (1997). The distance used for scaling disparities is the same as the one used for scaling retinal size. *Vis. Res.*, 37:757–764.

Dosher, B. A., Sperling, G., and Wurst, S. (1986). Tradeoffs between stereopsis and proximity luminance covariance as determinants of perceived 3D structure. *Vis. Res.*, 26:973–990.

Econopouly, J. (1995). Binocular disparities and kde are combined to rescale binocular disparities. Unpublished doctoral dissertation.

Econopouly, J. C., and Landy, M. S. (1995). Stereo and motion combined rescale stereo. *Invest. Ophthalmol. Vis. Sci. Suppl.*, 36:S665.

Foley, J. M. (1980). Binocular distance perception. *Psych. Rev.*, 87:411–434.

Gogel, W. G., and Tietz, J. D. (1973). Absolute motion parallax and the specific distance tendency. *Percept. Psychophys.*, 13:284–292.

von Helmholtz, H. (1910/1925). *Treatise on Physiological Optics*. New York: Dover.

Johnston, E. B. (1991). Systematic distortions of shape from stereopsis. *Vis. Res.*, 31:1351–1360.

Johnston, E. B., Cumming, B. G., and Landy, M. S. (1994). Integration of stereopsis and motion shape cues. *Vis. Res.*, 34:2259–2275.

Johnston, E. B., Cumming, B. G., and Parker, A. J. (1993). Integration of depth modules: Stereopsis and texture. *Vis. Res.*, 33:813–826.

Kaufman, L. (1974). *Sight and Mind*. New York: Oxford.

Landy, M. S., and Brenner, E. (1999). When does motion added to one object improve the judged shape of a nearby, static object? *Invest. Ophthal. Vis. Sci. Suppl.*, 40:S801.

Landy, M. S., Maloney, L. T., Johnston, E. B., and Young, M. J. (1995). Measurement and modeling of depth cue combination: in defense of weak fusion. *Vis. Res.*, 35:389–412.

Li, J. R., Maloney, L. T., and Landy, M. S. (1997). Combination of consistent and inconsistent depth cues. *Invest. Ophthal. Vis. Sci. Suppl.*, 38:S903.

Mamassian, P., and Landy, M. S. (1998). Observer biases in the 3d interpretation of line drawings. *Vis. Res.*, 38:2817–2832.

Mamassian, P., Landy, M. S., and Maloney, L. T. (in press). Bayesian modeling of visual perception. In R. P. N. Rao, B. A. Olshausen, and M. S. Lewicki (Eds.), *Statistical Models of the Brain*. Cambridge, MA: MIT Press.

Ono, H., and Comerford, J. (1977). Stereoscopic depth constancy. In W. Epstein (Ed.), *Stability and Constancy in Visual Perception: Mechanisms and Processes* (pp. 91–128). New York: Wiley.

Richards, W. (1985). Structure from stereo and motion. *J. Opt. Soc. Am. A*, 2:343–349.

van Rijn, L. J., and van den Berg, A. V. (1993). Binocular eye orientation during fixations: Listing's law extended to include eye vergence. *Vis. Res.*, 33:691–708.

Rogers, B. J., and Bradshaw, M. F. (1993). Vertical disparities, differential perspective and binocular stereopsis. *Nature* , 361:253–255.

Rogers, B. J., and Collett, T. S. (1989). The appearance of surfaces specified by motion parallax and binocular disparity. *Quart. J. Exp. Psych.*, 41A:697–717.

Sobel, E. C., and Collett, T. S. (1991). Does vertical disparity scale the perception of stereoscopic depth? *Proc. Roy. Soc. (Lond.) B*, 244:87–90.

Tittle, J. S., and Braunstein, M. L. (1990). Shape perception from binocular disparity and structure-from-motion. In P. S. Schenker (Ed.), *Sensor Fusion III: 3-D Perception and Recognition*, Volume 1383, (pp. 225–234).

Tittle, J. S., Todd, J. T., Perotti, V. J., and Norman, J. F. (1995). Systematic distortion of perceived three-dimensional structure from motion and binocular stereopsis. *J. Exp. Psych.: Human Percept. and Perf.*, 21:663–678.

Wallach, H., and O'Connell, D. N. (1953). The kinetic depth effect. *J. Exp. Psych.*, 45:205–217.

Wallach, H., and Zuckerman, C. (1963). The constancy of stereoscopic depth. *Am. J. Psych.*, 76:404–412.

Young, M. J., Landy, M. S., and Maloney, L. T. (1993). A perturbation analysis of depth perception from combinations of texture and motion cues. *Vis. Res.*, 33:2685–2696.

8

Signal Detection and Attention in Systems Governed By Multiplicative Noise

Christopher W. Tyler

Most signal detection analysis assumes additive sources of noise; attentional aspects are typically introduced in the form of uncertainty theory. This approach predicts that if your attention field is larger than the stimulus you are completely ignorant of where in the field the stimulus occurred. We have developed a new approach called distraction theory that is more plausible because it presumes accurate knowledge of location but distractability within the attention field. Concatenating either theory with the presence of multiplicative noise produces a range of unexpected results.

Theoretical papers by Shadlen, Newsome, and coworkers consider the implications of multiplicative neural noise on the role of individual neurons in the psychophysical performance of the organism as a whole. They reach the conclusion that linear summation over pools of 50–100 neurons is likely to provide the basis for behavior. Our analysis of 2AFC probability summation effects under multiplicative noise conditions suggests a profoundly different viewpoint in which nonlinear summation over small pools of neurons is a more plausible strategy to account for such performance.

8.1   Introduction

A key property of brain organization is the neural connectivity that provides access at each neuron to an extended range of input signals. The neural cell body generally acts as a linear summator, so the range of input signals to which the cell is responsive may be considered its summation field. The summation field (or "receptive field") of each neuron from the receptor level in the eyes provides for spatial-integration of visual information across the visual field. Despite previous attempts, a valid analytic framework has yet to be applied to the variety of spatial integration phenomena measured in laboratory studies. The analysis provided here forms the basis for a comprehensive theory of spatial summation based on the tenets of signal detection theory, specifically in the context of detection and discrimination tasks measured by the two-alternative forced-choice (2AFC) paradigm of the detectability of visual stimuli by neural signals anywhere in the brain. The analysis is valid for summation in any stimulus domain, but it will be illustrated with specific reference to summation in one- and two-dimensional spatial vision.

The kind of summation performed by physiological receptive fields will be termed *physiological summation* (whether linear or nonlinear) to distinguish it from *probability summation* performed on the outputs of a set of decision variables (even though this operation must also ultimately be a physiological process in the brain). The initial analysis will be developed under the assumptions that the main source of noise is external, Gaussian, and independent of stimulus contrast and that the detector has an ideal form in the sense that it is matched to the form of the stimulus. The theory is also sufficiently comprehensive to encompass conditions where threshold is dominated by internal Gaussian noise that is additively independent of the signal and can be shown to be relatively robust to the precise form of the noise distribution (Chen and Tyler, 1999). On the other hand, extension of the theory to cases where the noise is internal but varies with the strength of the neural signal reveals major departures from the independent noise behavior.

8.2 Signal Detection Theory for Ideal and Non-ideal Observers

8.2.1 Overview of ideal observer analysis

The ideal observer formalism assumes that the observer has complete knowledge of the stimulus and uses a single matched filter to detect its presence (Wiener, 1949). The ideal observer therefore is effectively a Bayesian detector with a prior probability of 1.0 on the matched filter and zero elsewhere. Optimal performance with this filter is assumed to occur with linear summation of the noisy filter inputs sampled by the field. As derived in the Appendix, the sensitivity for ideal detection improves with the square root of the number of local samples, or stimulus area.

Note, however, that there is a problem with applying this model in practice, because the psychometric function in this model is based on a linear relation between signal discriminability d' and signal strength. This linear relation is violated by most d' measurements, which typically show an exponent of about 2 (e.g., Stromeyer and Klein, 1974). Similarly, translation of this prediction into the Weibull format (Pelli, 1987) yields a predicted Weibull exponent of 1.3 (whereas most measurements show exponents of 3-4). Extension of the theory to non-ideal attention behavior, which encompasses steeper exponents, is left to the next section. First we consider an approximation to ideal behavior that can be used if the observer knows the set of stimulus types that may be presented in a block of trials, even if the particular stimulus is not known in advance on each trial.

If the task is summation over a range of stimulus sizes, the ideal observer model requires a summing receptive-field matching every size of stimulus for which the summation behavior is exhibited. A physiological implementation of such behavior

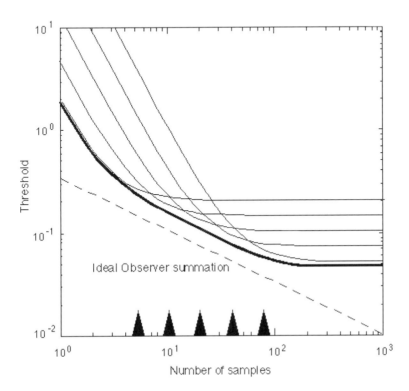

FIGURE 8.1. Physiological implementation of Ideal Observer behavior (dashed line). Thin curves: summation behavior for five individual Gaussian filters as the extent of a Gaussian test stimulus is varied (arrowheads indicate filter extents or number of input samples in each filter at half-height). The ideal strategy is to read out from that filter only when it matches the stimulus extent. At the two ends of the range, only one filter dominates detection behavior and hence system performance (thick line) departs from the ideal slope of -0.5 to follow the function for the most sensitive filter in that region. A fourth-power attentional summation operator (thick curve) also follows the envelope of the individual functions even though it has no information about the stimulus extent.

is depicted in Fig. 8.1, where the attention mechanism is assumed to switch to the receptive field size matching the stimulus presented in each condition. This behavior is possible only if the stimuli are presented in blocks of trials, so that the form of the next stimulus on each trial is known. If human observers exhibit a log-log summation slope of $-1/2$ (dashed curve in Fig. 8.1), they may be said to manifest ideal observer behavior, in the sense of using ideal matched filters to improve in the way an ideal observer would, even if the absolute sensitivity is less than predicted for an Ideal Observer (i.e., lower than ideal efficiency). Such (inefficient) Ideal Observer behavior may be taken as evidence that the brain has access to summing fields matching the sizes of all the tested stimuli, either present and selectable by attention, as in the central region of Fig. 8.1, or alternatively as an adaptive mechanism reforming itself for each new stimulus condition.

If it has access to only a limited range of sizes, the summation slope should asymptote to -1 for stimulus sizes below that of the smallest summing field size and should asymptote to 0 for sizes above that of the largest summing field, as depicted by the lower bound of the thin curves in Fig. 8.1. Thus, the form of the summation function in any stimulus domain carries important information about the range of summing field sizes operating in that domain (see Gorea and Tyler, 1997, for an example in the temporal domain and Kersten, 1984, for an example in the spatial domain). The model that the brain contains an adaptive filter reforming itself for each new stimulus condition seems to be incompatible with the occurrence of a limited summation range, because it is unclear why such adaptive capability would *fail* at a particular point.

8.2.2   The concept of probability summation

Probability summation corresponds to a decision rule where the response is defined by a signal being received from any one of a set of inputs. Its implementation requires a decision mechanism with access to a number of independent signals reflecting the occurrence of a stimulus. Although probability summation is often considered as a purely mathematical operation, the max operator, it is meaningless in the context of the human vision (in a single brain) unless the rule is mediated by some neural hardware. The neural implementation of such a decision rule may be designated as "attentional summation." This raises the issues of whether there are independent neural channels among which decisions may be selected and what is meant by a max operator in a neural system. In terms of detection theory, two channels are considered independent when they are governed by sources of noise that are statistically independent. There is plenty of evidence for a high degree of statistical independence among even neighboring cortical neurons (e.g., Freeman, 1994, 1996; Shadlen and Newsome, 1998), so cortical neurons can be considered to be separate channels for this purpose. To implement a max operator, a neural system must transmit a spike that initiated a detection response on receiving a spike from any one of its inputs. The threshold characteristic of cortical neurons with inputs weighted according to their receptive field thus provides the requisite hardware for a max operator.

In terms of the detection of signals in additive noise, the optimal strategy is to use a linear filter exactly matching the stimulus profile. It is possible to approximate ideal observer strategy with a neural system that performs attentional summation over the full set of filters in the form of the max of the signal-to-noise ratios (Pelli, 1985). This approach may be considered an ideal (or Bayesian) *attentional* strategy in that the observer knows the set of likely filters to survey on each trial and excludes all other filters. This strategy will have the effect of isolating the most efficient filter under any condition and hence will mimic ideal observer behavior without requiring prior knowledge of the particular stimulus.

It is common practice to combine the response outputs in neural-network models by a Minkowski summation rule:

$$R = \left[\sum_n (R_i^p) \right]^{\frac{1}{p}} \tag{8.1}$$

where the summing exponent is often set at $p = 4$. Note that such fourth-power summation (thick curve in Fig. 8.1) produces a completely smooth curve in the range where the discrete filters are present even though the assumed filters are separated in size by factors of 2. It is thus possible to approximate Ideal Observer behavior with relatively coarse physiological sampling in a particular domain if there is some way in the cortex to implement the Minkowski summation of Eq. (8.1) with a high summation exponent.

However, implementation of this strategy does require the neural system to have an accurate representation of the noise level, in order to compute the signal-to-noise ratios. Simply taking the max over raw signals will tend to emphasize the noisiest fields. But if it is plausible that the neural system normalizes to the prevailing (long-term) noise level, then a max operator would provide a mechanism for approximating ideal observer behavior. If the max operator has access to only a limited range of summation field sizes in the likely range, as shown as the thick curve in Fig. 8.1, the summation slope should match the form of the most sensitive summation function for each stimulus extent. Thus, the slope of the function should asymptote to -1 for stimulus sizes below that of the smallest summing field size (left-hand side of Fig. 8.1) and asymptote to 0 for sizes above that of the largest summing field (right-hand side of Fig. 8.1). Under the attentional strategy, the form of the summation function again reveals the range of available summing field sizes.

8.2.3 Attentional summation in 2AFC experiments derives from the difference distribution

For 2AFC detection using more than one channel, attentional (or 'probability') summation effects should be analyzed through signal detection theory. For a tractable analysis, we assume n stimulated channels of equal sensitivity with additive Gaussian noise. For the full analysis, we will consider the situation where the observing system is monitoring more channels (m) than are being stimulated. The statistical combination rule for attentional summation of the responses over channels is derived again from the maximum value of the set of m monitored channel responses in each stimulus interval (Pelli, 1985; Palmer, Ames, and Lindsey, 1993). For the null stimulus of the pair, which by definition contains only noise, the combined response distribution is based on the noise-alone distributions in the responses of all m channels. Mathematically, this combined distribution is given in terms of the expected values of the distributions by the derivatives, omitting the distribution variables for clarity:

$$
\begin{aligned}
M_m(R, \sigma_R) = \max_{i=1:m}[D_N] \quad &= \frac{d}{dr}\left[\int_{-\infty}^{r} D_N dr'\right]^m \\
&= mD_N\left[\int_{-\infty}^{r} D_N dr'\right]^{(m-1)}
\end{aligned}
\tag{8.2}
$$

The two parameters in the expression $M_m(R, \sigma_R)$ for the max distribution imply that we are deriving the form of the expected function of the resulting probability distribution, which may be characterized by the parameters of its location and spread.

With the inclusion of n signal channels for the signal interval of the stimulus pair, the max must be taken over the maxes of the separate n signal+noise and $m - n$ noise-alone distributions:

$$
\begin{aligned}
M_{n,m}(R, \sigma_R) &= \max\left[\max_{i=1:n}[D_R(r_i)]\max_{i=n+1:m}[D_N(r_i)]\right] \\
&= \frac{d}{dr}\left(\left[\int_{-\infty}^{r} D_R dr'\right]^n \cdot \left[\int_{-\infty}^{r} D_N dr'\right]^{m-n}\right)
\end{aligned}
\tag{8.3}
$$

In the general case, Eq. (8.3) does not simplify in the manner of eq. (8.2).

The simplest case of 2AFC attentional summation is the case where $m = n$, so there is no uncertainty as to which of the monitored channels contain the stimulus, and the two distributions differ only in their mean level of internal response. This situation corresponds to assuming an ideal *attention window* that always matches the stimulus extent, so that no unstimulated channels are monitored. Nevertheless, it is assumed that the observer cannot perform ideal *summation* over the stimulus area, but is forced to monitor the set of n local channels to find which gives the max response in any test interval (Pelli, 1985).

Figure 8.2 shows the numerical distributions for samples of maxes computed according to the derivation of Eq. (8.2) for noise alone (or eq. 8.3 for signal + noise with $m = n$) in factors of 10 from $n = 1$ to 1 million channels of equal sensitivity. The σ of these max distributions decreases by a factor of about 4. In each case, the observer's task is to distinguish between sample stimuli drawn from the max distributions of noise-alone and signal+noise for summation over a given number of channels. The signal+noise response has the same distribution as the noise-alone response (assuming all stimulated channels are monitored), except that it is shifted further rightward by the mean signal amplitude. Discriminability from Eq. (8.9) therefore improves with the reciprocal of the reduction in σ in these max distributions (Fig. 8.2), as shown for a full range of signal levels in the d' functions

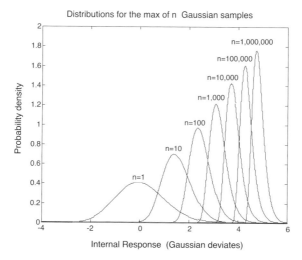

FIGURE 8.2. Max distributions for a Gaussian probability density function for numbers of samples increasing in factors of 10 from $n = 1$ to 1 million. Note the decreasing standard deviation and small asymmetry of these max distributions.

of Fig. 8.3a, shifted leftward in proportion to the reduction in σ. The consequent improvement in sensitivity measured at the point where each curve intersects the criterion level of $d' = 1$ is depicted in Fig. 8.3b (see Chen and Tyler, 1999, for details).

Thus, the complete analysis of 2AFC attentional summation over channels of equal sensitivity shows that the ideal attention operator provides dramatically different "probability summation" behavior than that implied by Pelli's (1985) high-uncertainty approximation to high-threshold theory, where the log slope was always -0.25. At its steepest, this 2AFC function reaches a slope of -0.25 (from 1 to 4 samples) but soon produces negligible summation for larger numbers of samples. The key reason for the difference between this prediction and that for the Weibull approximation is that the tails in the Gaussian distribution assumed here fall much more rapidly than the exponential tail of the Weibull distribution. A summation mechanism focusing on the information in this tail region will necessarily give different results for the two distributions.

Consider the practical implication of the summation function of Fig. 8.3b. For most reported neural tasks, the smallest stimulus employed might plausibly stimulate many local mechanisms. The expected starting point for a probability summation prediction would then be some way down this curve, say at the 10^2 level, beyond which little improvement is evident. Under the ideal attention assumption, the only way to achieve summation exponents even close to the reported psychophysical values of around -0.25 (Watson, 1979; Robson and Graham, 1981; Williams and Wilson, 1983; Pelli, 1985) would be to assume that attention can be focused onto a single neural channel for the smallest stimulus in the series.

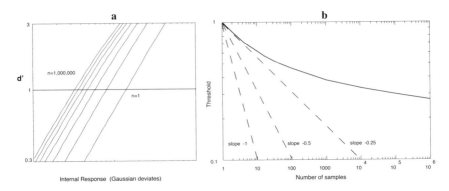

FIGURE 8.3. (a) Theoretical d' functions under 2AFC probability summation assumptions. Note that the exponent (or steepness) is almost invariant with number of equally sensitive channels monitored from $n = 1$ to 1 million (assuming no uncertainty). 2AFC summation behavior is therefore essentially invariant with the d' criterion selected. (b) 2AFC probability summation over six decades on (unequal) double-log coordinates, compared with summation slopes for full summation (-1), for ideal observer summation (-0.5) and for Weibull summation assuming $\beta = 4$ (-0.25). Note that the 2AFC summation function is never steeper than a slope of -0.25, and becomes extremely shallow for more than about ten samples.

8.2.4   2AFC attentional summation with uncertainty within a fixed attention window

Channel uncertainty theory is an elaboration of signal detection theory in which the number of neural channels m monitored in the brain is greater than the number of channels n stimulated (by ratio $M = m/n$) (derived formally in Pelli, 1985). Here we evaluate the particular case of the effect of uncertainty on threshold of varying stimulus extent with a fixed attention window (Fig. 8.4). For studies that do not expend the effort required to measure the psychometric slope, it is important to have a model of the effects of uncertainty under plausible assumptions. Clearly, if the attention window can be matched to the stimulus extent, the uncertainty will remain constant at zero and have no effect on the measured summation function. However, in this case the slope of the psychometric function should be low (assuming a linear transducer), which is known to be invalid in many situations.

8.3   Distraction Theory

A disturbing property of uncertainty theory is that it postulates that the neural system is incapable of resolving the location of any of the responses within the attention field. While this may seem plausible for non-human brains, it violates our experience in the psychophysical-detection task, which is that even in peripheral vision we have a pretty good idea of the location of the target when it was detected.

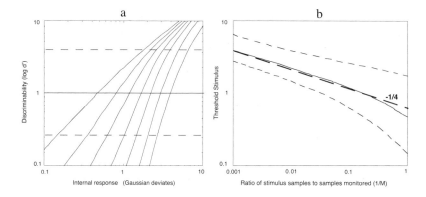

FIGURE 8.4. Probability summation for varying numbers of samples within a fixed attention window (assumed here to allow a maximum of 1000 samples). a) Psychometric functions in log d' vs. log stimulus strength. Note similarity in shape to those in Fig. 8.3a but with extra shifts at high uncertainties. b) Summation as a function of ratio of number of samples to total number monitored, at the three d' criteria indicated by the horizontal lines in (a). Thick dashed line in (b) depicts a slope of $-1/4$, which provides a good approximation to fixed-window probability summation over most of the computed range.

For example, in the automated visual-field tests performed for an eye examination, uncertainty is very high because the flash may appear anywhere throughout the visual field. But when we detect the flash, we have the impression that we could point to it with fair accuracy (unless we move our eyes, which will severely distort the spatial metric).

To accommodate the experience of good localizability, Kontsevich and Tyler (1999a) proposed the alternative of distraction theory. This approach assumes that the observer has a focal attention mechanism with a field of operation within which it continually shifts to the most salient stimulus. The observer is assumed to have accurate knowledge of all locations in the visual field but to be able to pay attention to only one location at a time. The uncertainty factor arises from the distracting effect of noise, which may exceed the stimulus in salience at some location when the stimulus is near threshold. In this case, attention is distracted to the wrong location and the observer is unable to process the stimulus at its presented location. The observer is presumed to be aware of being distracted, but by the time attention is redirected to the correct location, the stimulus has gone. Distraction Theory thus provides increasing opportunities for distraction as the attention field increases, just as does Uncertainty Theory, but without assuming any uncertainty.

Quantitative modeling of the effects of distraction within a defined attention field (Kontsevich and Tyler, 1999a) reveals that they are almost identical to those for uncertainty theory within the same field of attention. The acceleration of the psychometric function and the degree of attentional summation are statistically indistinguishable under the two theories. Informal observation supports the idea that the observer does, in fact, know the location of peripheral distractors to relatively high accuracy under conditions where the attention field is forced to be large, but we have not yet completed formal experiments to this effect.

8.4   Effects of Multiplicative Noise

8.4.1   Multiplicative noise makes the psychometric function shallower

Instead of the classical assumption of additive noise, analysis of the noise in cortical neurons suggests that it may have a multiplicative component, with σ_R increasing according to some function of the strength of the mean signal R (Tolhurst, Movshon and Thompson, 1981). A general expression for the total noise in the signal is the power relation:

$$\sigma_R \propto kR^q + \sigma_N \tag{8.4}$$

The additive constant σ_N represents some irreducible level of noise that is present even when there is no signal, when the multiplicative component kR^q will fall to zero. Such additive noise is a physiological requisite because no real system is noise-free.

8.4.2   Dramatic probability summation with multiplicative noise

If we evaluate the effects on sensitivity of taking the max of n equally-sensitive channels in the presence of root-multiplicative Gaussian noise, the results are also profound (Fig. 8.5b). Summation over the first ten channels actually exceeds the amount expected for linear summation in additive noise. This result seems counterintuitive, but it arises because any decrease in the signal provides a concomitant decrease in the accompanying noise level. Becuase the effect of probability summation is to allow a small decrease in the signal at the outset, the multiplicative reduction in the noise provides a further enhancement of the signal/noise ratio, resulting in the net improvement depicted in Fig. 8.5b.

This dramatic degree of probability summation under root-multiplicative noise conditions has powerful implications for neural processing, which is governed by this type of principle throughout the cortex (Tolhurst, Movshon and Thompson, 1981; Tolhurst, Movshon and Dean, 1983; Vogels, Spileers and Orban, 1989). Shadlen and Newsome (1998) point out that the multiplicative behavior makes the signals at individual neurons so noisy that they cannot account for the discriminative behavior of the animal as a whole, even if the neuron's response is optimal for the local stimulation employed. They estimate that the activity must be integrated over 50-100 neurons to account for the observed behavior, implying that the signal/noise ratio of the optimal neuron is about a log unit below the required level.

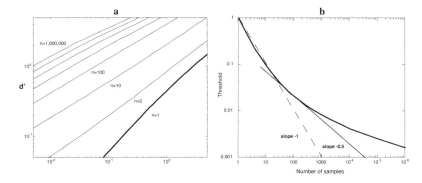

FIGURE 8.5. Effects on probability summation of assuming square-root multiplicative noise according to Eq. (8.4) with $p = 0.5$. (a) Shallower d' functions. (b) Dramatically enhanced summation behavior for threshold stimulation at the criterion of $d' = 1$ that is even supralinear for small numbers.

However, the plot of Fig. 8.5b implies that a different strategy is available under root-multiplicative noise conditions. Instead of integrating the activity of 100 neurons, and losing the potential specificity available from the elements of that assemblage, the cortex could monitor the activity of just ten *relevant* neurons. Taking the *maximum* of the ten responses gives the required boost of a factor of ten in signal/noise ratio, equivalent to *summing* over one hundred neurons. Thus, a much smaller pool is required for the same gain in detectability, if the brain is capable of implementing a maximum-detecting rule. Such implementation seems plausible because it is the core operation of an attentional process, for which there is much behavioral and increasing neurophysiological evidence. In fact, a simple neural threshold has the effect of implementing a max rule in a psychophysical task where the stimulus is reduced until the last response of the most sensitive neuron carries it.[1]

8.4.3   Suprabehavioral neural sensitivity and its implications for attentional selection

The result reported by Shadlen and Newsome (1998) – that the signals at individual neurons are too noisy to account for the discriminative behavior of the animal as a whole – is based on the *average* response of the neurons that they recorded in the MT region to a motion stimulus. However, they also point out that the most sensitive neuron they recorded was as much as 10 times *more* sensitive than the animal's behavior. Both sensitivities were measured in the commensurate units of d', the statistical discriminability carried by the respective noisy signals. This astounding result implies that, if the monkey had had access to the signal from the best available

---

[1]The detailed effects of a hard threshold on 2AFC performance, which are beyond the scope of the present treatment, are discussed in Kontsevich and Tyler (1999b).

neuron, it would have been able to improve its behavioral discrimination by a factor of 10. The fact that it was not able to capitalize on this signal leads to a number of surprising conclusions about neural organization.

1. The information available at single neurons in the MT area is not generally available to guide the response of a trained and motivated animal. Whatever the neural organization underlying this disconnect, is seems a curiously suboptimal way for the nervous system to have evolved.

2. If we assume that the animal was fully trained, there are two possibilities: either the attention mechanism is guided by a variable other than the detection efficiency of the signals from single neurons, or the attention mechanism does not have access to single-neuron signals in this cortical area.

3. If attention is not driven by the detection efficiency, the most likely alternative is that it is driven by signal strength. The implication would therefore be that neurons with the same signal strength vary in signal to noise ratio by a factor of about 100 (based on the range of d' reported by Shadlen and Newsome (1998). In this scenario, the attention mechanism would be attracted to the largest signal and could miss a signal with 10 times the detection efficiency because the signal strength was smaller.

4. The other main alternative is that signals from single neurons are unavailable because they are a local component of a larger summing field that has the function of integrating global properties of the motion field impinging on the animal. Attention (in humans, we would say "consciousness") would have access only to the output of this global field, not the local computations contributing to it. On an optimal scheme for neural specialization, there would be another brain area specialized for targeting the local signals, which would then be fed to the global processing in MT. On this hypothesis, the Shadlen and Newsome recordings in MT would represent the input signals from this earlier brain area, while all output neurons in MT would be engaged in global summation computations of various kinds. However, one would have expected the high discriminability of the local signals to be accessible to behavior from this earlier processing level (e.g., V2) even if it were inaccessible in MT. But the monkey's behavior implies that this was not the case.

5. One can infer strongly that the monkey does not have available to it a max operator capable of extracting the maximum signal to noise ratio of a large set of MT signals from the type of task used by Shadlen and Newsome (1998). Even though theirs was a suprathreshold noise task, the max operator may be straightforwardly implemented by introducing an accelerating nonlinearity before averaging (Minkowski summation). The sum will then be dominated by the largest of the signals. It would not, however, extract the largest signal to noise ratio unless the noise levels were similar in all neurons.

6. The assumption of Shadlen and Newsome (1998) that the system would necessarily integrate over 50–100 neurons is somewhat implausible because it
   implies not only massive redundancy but also the loss of hard-won information about the world. Why average over the neural pool to lose a factor
   of 10 in information when you could perform a max operation and gain a
   factor of 10?

7. Shadlen and Newsome's task was a suprathreshold discrimination of motion direction in a noise field. The alternative to the assumption of linear
   integration over 50–100 neurons is to assume that the attention mechanism
   is designed to operate as a Minkowski operator near threshold with an accelerating nonlinearity. Any accelerating nonlinearity tends to be linearized
   by the presence of noise, converting the summation to a linear process in
   this suprathreshold regime. The attentional summing mechanism will then
   fail to select the max and will operate like a linear integrator, as proposed
   by Shadlen and Newsome.

8. On this hypothesis, the behavior would be very different in a contrast-
   threshold task. Cortical neurons have a built-in threshold, gating out the
   high level of noise in the signals reaching the cortex from the lateral geniculate nucleus. These thresholds, or accelerating nonlinearities, convert the
   summation into a max operator, allowing only the strongest signals to reach
   the cortex when stimulus contrast is low. When the stimulus is "at threshold," the strongest signal from any of the monitored neurons is a fortiori
   the only signal. If attention is drawn to the strongest (i.e., the only) signal,
   it will then select the correct 2AFC interval even if it cannot be directed to
   the relevant neuron.

8.4.4   Fully multiplicative noise introduces psychometric saturation

A more extreme form of multiplicative noise is the case where the noise is directly
proportional to the stimulus strength ($q = 1$ in Eq. (8.4)). Direct proportionality is
not implausible, as such a form occurs in the case of noise due to eye-movement
fluctuations over the image being viewed. Whatever form the distribution of eye
movements takes, it will introduce some level of fluctuation in the response of
a local filter viewing any kind of contrast stimulus. The resulting fluctuation is
a form of noise that is necessarily in direct proportion to the stimulus contrast
(assuming that the eye movements are independent of contrast). This property of
direct proportionality may be shown analytically in terms of the temporal waveform
of the signal fluctuation of the output of each linear filter $I_{k_i}(x, y)$ responding to
some stimulus $S(x, y)$, such as a sinusoidal grating, projected onto the moving
retina.

$$\begin{aligned} r_i(t) &= I_{k_i}(x, y) \otimes s \cdot S(x - \Delta x(t), y - \Delta y(t)) \\ &= s \cdot [I_{k_i}(x, y) \otimes S(x - \Delta x(t), y - \Delta y(t))] \end{aligned} \tag{8.5}$$

where $\otimes$ is the convolution operator, $\Delta(x)$, $\Delta(y)$ is the retinal shift over time, and s is the scaling constant of stimulus strength.

Thus, for a given filter and eye-movement sequence, the filter outputs $r_i(t)$ are directly proportional to the contrast of the stimulus because convolution is a linear operation. We may treat the response of each filter derived from such eye movements as a noise source by determining its standard deviation σ_E computed over some temporal window $t_1 : t_2$ according to:

$$\sigma_E = \left[\int_{t_1}^{t_2} \left(r_i(t) - \int_{t_1}^{t_2} r_i(t)dt \right)^2 dt \right]^{\frac{1}{2}} \propto r_i(t) \propto s \qquad (8.6)$$

The standard deviation of this source of noise is thus directly proportional to the contrast of the background display. Such proportional noise will tend to overtake other sources of noise that do not increase so rapidly with contrast, and will therefore tend to dominate at high contrast.

The effect of proportional multiplicative noise ($q = 1$ in Eq. (8.4)) on the form of the psychometric functions is shown in Fig. 8.6. The fitted slopes become even shallower than in the case of square-root noise (Fig. 8.5a) when the signal rises out of the additive-noise regime where the slopes approximate unity. As stimulus strength increases, the effect of the multiplicative noise is to make the functions asymptote to a constant d' level, with no further improvement in sensitivity at high stimulus strengths. This horizontal asymptote thus becomes a conspicuous signature of the presence of fully multiplicative noise. Such behavior has rarely been seen in psychometric functions for contrast detection (e.g., it is not evident in the high-contrast study of Foley and Legge, 1981), suggesting that this type of multiplicative noise is not a usual feature of contrast detection tasks. However, it is not clear that previous workers have designed their studies for careful evaluation of this high d' region of the psychometric function, so there is room for further evaluation of particular situations of interest before the case may be considered to be settled. For example, although such noise might not be expected in a simple detection task, it might plausibly be found in a difficult discrimination task where the contrast threshold is being measured as a function of a slight spatial difference between two stimuli, with long-duration presentations allowing eye-movement-generated noise in the pedestal stimulus to become a significant factor limiting discrimination performance.

For summation over increasing numbers of mechanisms, the curves of Fig. 8.6 show that the additive noise regime (approximating a slope of 1) tends to dominate the domain of measurable (and computable) range of d' functions. As a result, derivation of a summation curve for this case is relatively meaningless because its form would depend on the exact ratio of additive to multiplicative noise assumed. When in the domain dominated by fully multiplicative noise (horizontal leg of curves), summation is indeterminate because reduction of the signal would be accompanied by a proportionate reduction of the noise, and signal to noise ratio (discriminability) would be maintained at a constant level. In the fully multiplicative noise regime, therefore, discriminability is insensitive to the signal level,

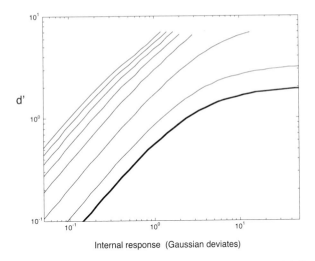

FIGURE 8.6. Saturating d' functions obtained assuming linearly multiplicative noise according to Eq. (8.6) with $q = 1$. Thick curve: one mechanism, next curve to left: two mechanisms, successive leftward curves from 10 to 1,000,000 in factors of 10.

and threshold cannot be determined. Only when the signal level is finally reduced into the domain dominated by additive noise (left-hand side of Fig. 8.6) would summation revert toward the form depicted in Fig. 8.3b.

Appendix: Derivation of Square-Root Summation for the Ideal Observer

Optimal performance with each filter in a bank of stimulus-matched filters is assumed to occur with linear summation over the noisy filter inputs l_j sampled by the field. The summation properties of the filters will vary with respect to a large number of stimulus attributes. For simplicity, we consider the case of spatial summation over two-dimensional stimuli varying in one dimension of overall size, k. This variable size dimension could be the height, the width, the area, or any parameter that is linear with the number of sources of input to each summing field i, indexed by $j \in l : k_i$. When the local regions have identical sources of independent Gaussian noise, the summed output of each field r_i is given by:

$$r_i = \sum_x \sum_y I_i(x, y) \cdot l_j, \quad \text{where } x, y \in -\infty : \infty \quad (8.7)$$

Note that j is specific to each field by the i subscript which is omitted from this index for simplicity. Expressed in terms of the mean values of the random variables $\bar{l}_j$, we see that the mean integrated response R_i is proportional to the effective area

a_i of each field:

$$R_i = a_i \cdot \bar{l}_j, \quad \text{where } a_i = \sum_x \sum_y I_i(x, y) \tag{8.8}$$

Since the ideal matched filter $l_{k_i}(x, y)$ increases directly with the effective area of stimulus i, the signal will increase proportionately while the noise should increase according to the Gaussian summation rule of the square root of the number of sources included, k_i (as derived from the effective area, a_i, times the source density). Thus, overall signal/noise ratio for a set of matched filters processing sets of signals of equal strength should improve according to:

$$d_i' \propto \frac{R_i}{\sigma_{R_i}} \propto \left[\frac{\sum_j \bar{l}_j}{\left(\sum_j \bar{l}_j^2 \right)^{\frac{1}{2}}} \right] \propto k_i^{\frac{1}{2}} \tag{8.9}$$

Thus, the sensitivity for ideal detection improves with the square root of number of local samples, or stimulus area.

Acknowledgments

Thanks to Chien-Chung Chen for performing the computational analysis and Lenny Kontsevich and Mike Shadlen for cogent discussions. Supported by NIH grant EY 7890.

References

Chen, C.-C. and Tyler, C. W. (1999). Accurate approximation to the extreme order statistics of Gaussian samples. *Comm. in Statistics: Sim. and Comput.*, 28:177-188.

Foley, J. M., and Legge, G. E. (1981) Contrast detection and near-threshold discrimination in human vision. *Vis. Res.*, 21:1041-1053.

Gorea, A., and Tyler, C. W. (1997) New look at Bloch's law for contrast. *J. Opt. Soc. Am. A*, 14:52-61.

Kersten, D. (1984) Spatial summation in visual noise. *Vis. Res.* 24:1977-1990.

Kontsevich, L. L. and Tyler, C. W. (1999a). Distraction of attention and the slope of the psychometric function. *J. Opt. Soc. Am. A*, 16:217-222.

Kontsevich, L. L., and Tyler, C. W. (1999b). Nonlinearities of near-threshold contrast transduction. *Vis. Res.*, 39:1869-1880

Palmer, J., Ames, C. T., and Lindsey, D. Y. (1993). Measuring the effect of attention on simple visual search. *J. Exp. Psych.: Human Percept. and Perf.*, 19:108-130.

Pelli, D. (1985). Uncertainty explains many aspects of visual contrast detection and discrimination. *J. Opt. Soc. Am. A*, 2:1508-1531.

Pelli, D. (1987). On the relation between summation and facilitation. *Vis. Res.*, 27:119

Robson, J. G. and Graham, N. (1981). Probability summation and regional variation in contrast sensitivity across the visual field. *Vis. Res.*, 21: 409-418.

Shadlen, M. N. and Newsome, W. T. (1998). The variable discharge of cortical neurons: implications for connectivity, computation and information coding. *J. Neurosci.*, 18:3870-3896.

Stromeyer, C. F. III and Klein, S. A. (1974). Spatial frequency channels in human vision as asymmetric (edge) mechanisms. *Vis. Res.*, 14:1409-1420.

Tolhurst, D. J., Movshon, J. A., and Dean, A. F. (1983). The statistical reliability of signals in single neurons in cat and monkey visual cortex. *Vis. Res.*, 23: 775-785.

Tolhurst, D. J, Movshon, J. A., and Thompson, I. D. (1981). The dependence of response amplitude and variance of cat visual cortical neurones on stimulus contrast. *Exp. Brain Res.*, 41:414-419.

Vogels, R., Spileers, W., and Orban, G. A. (1989). The response variability of striate cortical neurons in the behaving monkey. *Exp. Brain Res.*, 77:432-436.

Watson, A. B. (1979). Probability summation over time. *Vis. Res.*, 19:515-522.

Wiener, N. (1949). *Extrapolation, Interpolation and Smoothing of Stationary Time Series, with Engineering Applications*. Cambridge, MA: MIT Press.

Williams, D. W. and Wilson, H. R. (1983). Spatial frequency adaptation affects spatial probability summation. *J. Opt. Soc. Am.*, 73:1367-1371.

9

Change Blindness: Implications for the Nature of Visual Attention

Ronald A. Rensink

In the not-too-distant past, vision was often said to involve three levels of processing: a low level concerned with descriptions of the geometric and photometric properties of the image, a high level concerned with abstract knowledge of the physical and semantic properties of the world, and a middle level concerned with anything not handled by the other two.[1]

The negative definition of mid-level vision contained in this description reflected a rather large gap in our understanding of visual processing: How could the here-and-now descriptions of the low levels combine with the atemporal knowledge of the high levels to produce our perception of the surrounding world?

A number of experimental and theoretical efforts have been made over the past few decades to solve this "mid-level crisis." One of the more recent of these is based on the phenomenon of *change blindness*—the difficulty in seeing a large change in a scene when the transients accompanying that change no longer convey information about its location (Rensink, O'Regan, and Clark, 1997; Rensink, 2000a). Phenomenologically, this effect is quite striking: the change typically is not seen for several seconds, after which it suddenly snaps into awareness.[2] During the time the change remains "invisible," there is an apparent disconnection of the low-level descriptions (which respond to the change) from subjective visual experience (which does not). As such, this effect would seem to have the potential to help us understand how mid-level mechanisms might knit low- and high-level processes into a coherent representation of our surroundings.

---

[1]More precisely, low-level vision determines the scene-based properties (e.g., surface color and orientation) that give rise to the pattern of illumination on the retina, separating out the effects of extraneous factors such as lighting, occlusion, and noise. The result is a retinotopic sketch containing a detailed description of the visible scene at that moment in time (Marr, 1982). High-level vision, in contrast, involves issues of meaning. Among other things, it involves knowledge of object types, both in regard to how each type relates to the others and how each appears visually (e.g., automobile tires are almost always black and have an outer perimeter that is approximately round). Mid-level vision must somehow describe the scene using high-level knowledge about the types of object present, and information about particular parameters (e.g., location, time, size) obtained from the low-level descriptions.

[2]To experience this effect, see the QuickTime examples on the accompanying CD-ROM. Examples can also be found at http://www.cbr.com/projects.

It is argued here that this potential can indeed be realized and that change blindness can teach us much about the nature of mid-level vision.[3] A number of studies are first reviewed showing that the perception of a scene does not involve a steady buildup of detailed representation—rather, it is a dynamic process, with focused attention playing one of the main roles, namely, forming coherent object representations whenever needed. It is then argued that change blindness can also shed considerable light on the nature of focused attention itself, such as its speed, capacity, selectivity, and ability to bind together visual properties into coherent structures.

9.1   Visual Attention: Role in Scene Perception

9.1.1   Change blindness

Change blindness can be defined as the induced failure of observers to detect large changes in a visual display (Rensink et al., 1997). Under normal viewing conditions it is usually easy to notice when an item suddenly changes its color, location, or other attribute. But if the motion transients that accompany the change cannot provide information about its location, even large changes can become difficult to detect.

Consider the flicker paradigm of Rensink et al. (1997), shown in Fig. 9.1. Here, an original image of a scene alternates with the same image modified in some way (e.g., an item changes color or is moved). If the images alternate without intervening blank fields, the change is easily seen. But if a brief blank (about 80 msec or more) is interposed between them, detection of the change is dramatically impeded. For example, detection of the change shown in Fig. 9.1 requires over 40 alternations on average, even though the change is large, made repeatedly, and the observers are actively searching for it. This does not depend on the particular change or the particular image—similar results have been found for different kinds of changes on a wide variety of images (Rensink et al., 1997). In addition, this effect is robust, occurring for a wide range of blank intervals and colors (Rensink, O'Regan and Clark, 2000; see Figure 9.2).

Indeed, change blindness can be induced by a large number of techniques, such as making the change during:

- an eye movement (Bridgeman, Hendry, and Stark, 1975; Grimes, 1996),

---

[3]It should be mentioned that there is at the moment some non-uniformity in terminology. Some authors (e.g., Ullman, 1996) use "high-level" to refer to object perception, leaving "mid-level" to refer to the perception of relatively unstructured surface properties. Others (e.g., Henderson and Hollingworth, 1999) use "high-level" to refer to the perception that includes meaning (i.e., identifying an object or a scene as something), with "mid-level" vision referring to the formation of particular objects. As is hopefully evident, usage here is similar to that of Henderson and Hollingworth: "mid-level" refers to the formation of coherent object representations, with the rapid interpretation of scene properties (i.e., the formation of proto-objects, surfaces, etc.) regarded as a secondary stage of low-level processing.

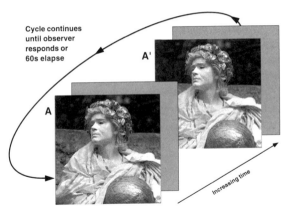

FIGURE 9.1. Example of flicker paradigm. Sequence alternates between original and modified image until observer responds. Display times (on times) are typically 200–600 ms, while blank times (off times) are 80–800 msec. In the stimulus here, original image A (statue with wall in background) and modified image A′ (statue with wall lowered) appear in the order A, A′, A, A′,... with gray fields placed between successive images. For this example, over 40 alternations are required on average before the observer detects the change.

- an eye blink (O'Regan, Deubel, Clark, and Rensink, 2000)

- a movie cut (Levin and Simons, 1997)

- occlusion of the changing item (Simons and Levin, 1998)

- small transient "splats" elsewhere in the image (Rensink et al., 2000).

(See Rensink, 2000c, or Simons and Levin, 1997 for a more complete review of various change-blindness studies.) Given that change blindness can be induced in many ways and that it has a strong phenomenological effect, it follows that change blindness is not an aberrant phenomenon occurring only under a special set of conditions. Rather, it appears to touch on something important – something central to the way that the world is perceived.

9.1.2   Coherence theory

Why is it that change blindness can be so easily induced? And if it is so easy to induce, why are we nevertheless so good at seeing changes in everyday life?

All the techniques used to induce change blindness share a comment element: the transients associated with the change are swamped (or otherwise neutralized) so that there is little information about its location in the display. This leads to the suggestion that *focused attention is necessary to see change* (Rensink et al., 1997). The world is by and large a quiet place—any change will likely be the only one occurring at that moment. As such, it will generate a motion signal in the image that will automatically attract attention to its location (see, e.g., Klein et al., 1992). However, if this signal is swamped by others occurring at the same

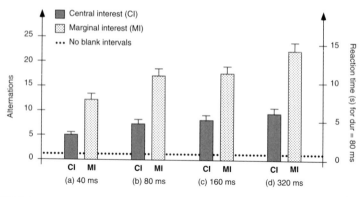

FIGURE 9.2. Time to detect change for various durations of blank intervals. Data from Rensink et al. (2000). CIs are central interests, items often mentioned in brief verbal descriptions of the scene; MIs are marginal interests, items never mentioned. (See Rensink et al, 1997, for more discussion of these terms.) Error bars indicate one standard error. The dotted line indicates performance when no blank intervals are present. (a) 40 msec durations. Although intervals were very brief, changes took far longer to detect than when no blanks were present. (b) 80 msec durations. Detection (analyzed in terms of number of alternations to see the change) took even longer than in the 40 msec condition. (c) 160 msec durations. No significant differences were found between this and the 80 msec condition. (d) 320 msec durations. Detection was reliably slower than for the other conditions, although the amount of slowdown was not large.

time, attention will not be automatically sent to the location of the change; instead, a more effortful attentional scan must be used. Because attention can operate on only a few items at a time (e.g., Pashler, 1988; Pylyshyn and Storm, 1988), and because there are many items in most real-world scenes, this scanning will usually take considerable time. The result is change blindness.

Note that this explanation requires that attention cannot "weld" visual features into a detailed, relatively long-lasting, coherent[4] representation (Kahneman et al., 1992). Indeed, it suggests that focused attention may endow a structure with a coherence that lasts only as long as attention is directed towards it. This viewpoint is given a more precise formulation as a *coherence theory* of focused attention (Rensink, 2000a):

- Prior to focused attention, structures are formed rapidly and in parallel across the visual field. These *proto-objects* can be quite complex, but have coherent structure only within a limited area of space (Rensink and Enns, 1995, 1998). Their temporal coherence is likewise limited—they are volatile, being constantly regenerated. As such, they are simply *replaced* by any new stimuli

---

[4]In this chapter, "coherent" refers to consistency and logical interconnection (i.e., agreement that the structures refer to parts of the same system). This term is used in two ways: in its spatial aspect, it denotes a set of representations at different locations that refer to the same object; in its temporal aspect, it denotes a set of representations at different times that refer to the same object (see Rensink, 2000a).

appearing in their retinal location.

- Focused attention acts as a metaphorical hand that grasps a small number of these proto-objects from the constantly regenerating flux. While held, these form an individuated object with a high degree of coherence over time and space. Such coherence is obtained via feedback between the proto-objects and a mid-level *nexus*, a locus where lower-level information is collected via a set of links. This allows the object to retain its identity across brief interruptions; as such, it is transformed rather than replaced by new stimuli arriving at its location.

- After focused attention is released, the object loses its coherence and dissolves back into its constituent set of proto-objects. This implies that there is little short-term memory apart from what is being attended. Such a position is also consistent with results indicating a lack of attentional aftereffect in visual search (see also Wolfe, 1999)[5].

This view of visual processing is illustrated in Figure 9.3. Note that the separation between low- and mid-level vision is a true divide, with both levels qualitatively different from each other. The low-level processes involve retinotopic representations with a considerable amount of visual detail,[6] but which are also volatile, lasting only a fraction of a second. Unattended proto-objects are therefore in constant flux, being built anew as long as light continues to enter the eyes. In contrast, the mid-level attentional processes involve a much sparser set of structures that are stable,[7] both in time (lasting as long as attention is directed to them) and in space (remaining invariant with eye movements).

According to coherence theory, then, low-level representations of considerable detail—proto-objects—exist as long as the scene continues to be projected to the eyes. Change blindness stems not from a lack of detailed representation but rather from an inability to make the entire set of proto-objects coherent enough to support the perception of change. It is important to note that although unattended proto-objects may be volatile and so cannot be directly reported, they can still provide

---

[5]In this view, focused attention is taken to be object-based rather than spaced-based. Part of the justification for this can be seen from the example in Fig. 9.1. Here, it takes considerable time before the change to the background wall is seen. If focused attention involved a spotlight of the type proposed in most space-based theories (e.g., Treisman and Gormican, 1988), some of it should "spill over" onto the wall while examining the statue, resulting in relatively fast detection of change. However, this does not occur, indicating that attention is allocated to relatively discrete structures (proto-objects) that correspond to various objects in the scene.

[6]The amount of detail in the representation will fall off with eccentricity from fixation. However, there is still considerable resolution even at eccentricities of several degrees (see e.g., Woodhouse and Barlow, 1982).

[7]In this chapter, "stable" is used in two ways: In its spatial aspect, it denotes invariance over eye movements; in its temporal aspect it denotes invariance over time, i.e., a representation that is not volatile. Note that both properties are required for a buffer that collects information into a representation independent of any particular viewing position.

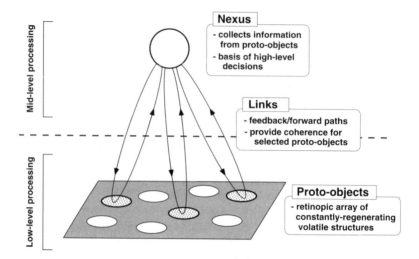

FIGURE 9.3. Relation of low- and mid-level processing. In the absence of focused attention, low-level structures (proto-objects) are volatile. Attention acts by establishing feedback links between proto-objects and a mid-level nexus. The set of interacting proto-objects, links, and nexus is a coherence field. The interaction among the various parts of the field allows the establishment of coherence in the properties of the selected proto-objects, both in space and in time.

an immediate context that influences the perception of the attended structures that are reported (Moore and Egeth, 1997).

9.1.3 Virtual representation

Change blindness shows that observers are poor at combining information from successive images—they can neither detect a difference directly, nor detect the superposition of items that would occur if information was accumulated. This indicates that there is no large-scale, detailed buffer into which information is collected. Indeed, coherence theory asserts that no more than a few coherent object representations exist at any time. But if this is so, how can a scene be represented? And why do we have the impression that we observe a large number of coherent objects simultaneously?

The answer to these questions centers around the idea of a *virtual representation*: instead of creating a detailed coherent representation of all the objects in the scene, do so only for those few objects needed for the task at hand. If a coherent representation of an object can be formed whenever requested, the resultant representation will appear to higher levels as if it is "real" (i.e., as if all objects simultaneously have a coherent representation). Such a scheme will have almost all the power of a complete set of object representations while requiring far less in the way of processing and memory resources (Rensink, 2000a).

The success of a virtual representation depends on its ability to form a coherent representation whenever requested. Fortunately, this is easy to do, at least in principle: when a request is made, shift attention and the eyes to the location of the object in the image; then, obtain detailed information from the incoming light and incorporate it into a coherent representation. Note that a high-capacity memory is not needed—detailed information about any object in the scene can almost always be obtained from the world itself. As pointed out long ago by Stroud (1955, p. 199):

> "Since our illumination is typically continuous sunlight and most of the scenery stays put, the physical object can serve as its own short-term memory...".

Provided that attention and eye movements are properly coordinated, the result will be a virtual representation whose dynamic nature is transparent to processes at higher levels.

Given this, the question remains of how such coordination might be carried out. One possibility is an architecture in which there exists not only an attentional stream for the processing of objects but also a concurrent nonattentional stream that forms stable structures for attentional guidance (Rensink, 2000a). The resulting system — shown in Fig. 9.4 — has three subsystems. Each of these obtains its input from low-level vision and is informed by the abstract knowledge available at high levels:

- a limited-capacity attentional system that forms low-level proto-objects into coherent, low-capacity object representations.

- a limited-capacity non-attentional system that uses the statistics of the proto-objects to determine the gist of the scene (i.e., its abstract meaning).

- a limited-capacity non-attentional system that uses the locations of the most significant proto-objects to determine the layout of objects in the scene.

In this architecture, attention retains the role assigned to it by coherence theory, namely, the temporary formation of coherent objects. What has now been added is a setting system that maintains both scene layout and gist, but involves little (if any) focused attention. The feasibility of such a system is suggested by several studies showing that the extraction—or at least the maintenance—of gist and layout information does not require attentional processing (see, e.g., Henderson and Hollingworth, 1999; Rensink, 2000a). In this view, the mid-level processes that link low- level structures with high-level knowledge do not form a unitary system. Rather, they form a heterogeneous collection of subsystems, each with its own particular characteristics.

Given this architecture, a relatively simple set of interactions could carry out the visual perception of the scene. First, low-level processes would provide a constantly-regenerating description of the scene-based properties visible to the viewer. A subset of these could determine scene gist and layout, which could then invoke high-level knowledge about the objects and events that might be expected.

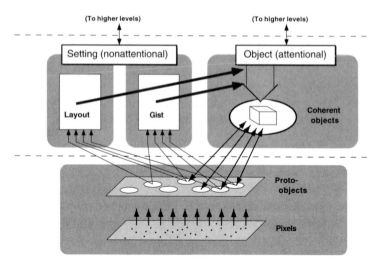

FIGURE 9.4. Architecture implementing virtual representation. Three different mid-level systems are involved: (i) an attentional system that gives selected proto-objects temporal and spatial coherence; (ii) a nonattentional setting system that uses low-level information to obtain the gist of the scene; (iii) a nonattentional setting system that uses low-level information to obtain the spatial layout of the objects in the scene. The latter two systems operate in tandem with the attentional system and provide information that helps guide it, so that it can form the appropriate coherent representation when requested.

These expectations could be tested via attention, which could provide detailed, coherent descriptions of selected objects, with the expected structure (and importance) of these objects being obtained from high-level knowledge. The perceived layout of the scene could facilitate this process, helping to guide attention to the appropriate low-level items.

Note that in this view of mid-level vision, focused attention still plays an important role, namely, the formation of coherent object representations. However the role of these representations has been reduced. Coherent representations are no longer needed for the recognition of various types of scenes—this can be largely done via the statistics of relatively simple properties. Indeed, coherent representations may not even be needed to recognize objects—at least in regard to object type. In other words, it may be possible to determine the presence of an object simply via the statistics of the properties in some part of the image. Coherent representations would then be primarily involved in describing particular instantiations of these types, assigning them temporal and spatial coordinates, as well as—via the links to the proto-objects—properties such as size and color. The formation of these kinds of descriptions would be necessary only for a more restricted set of operations, such as the intensive verification (scrutiny) of perceptual hypotheses about objects and scenes, and perhaps the selection and guidance of actions.

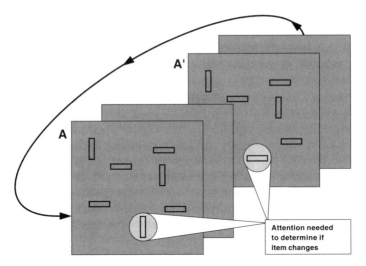

FIGURE 9.5. Example of flicker paradigm with controlled displays. Here, displays are arrays of rectangles. In half the displays, one item (the target) changes orientation while the others (the distractors) remain constant. According to coherence theory, the only way for an observer to determine if a change is occurring is to carry out an attentional scan of each item in the display.

9.2   Visual Attention: Mechanisms

9.2.1   Methodology

Change blindness not only provides a way to determine how focused attention relates to the rest of vision—it can also provide a way to explore the nature of attentional processing itself.

One possible way to do this is by replacing the images of scenes in the original flicker paradigm (Section 9.1) with arrays of simple figures (Rensink, 2000b). To see how this works, consider an array of rectangles, where one item (the target) changes its orientation on half the trials while the other items (the distractors) do not; the observer must then report for each trial whether or not a change was present (Fig. 9.5). If roughly half the items are horizontal and half vertical in both images, the target cannot be detected from any single display—both displays must be compared. If the interstimulus interval (ISI) between displays is sufficiently long, the transients due to the changing target will be swamped by the transients produced by the flickering distractors, requiring an attentional scan to carry out the task.

This approach therefore extends the "classic" visual search paradigm on static displays (e.g., Treisman and Gormican, 1988) into a more dynamic realm. All the power of the static techniques (e.g., investigating different shapes, different features, search asymmetries) is retained, while allowing manipulation of two more degrees of freedom: display time (on-time) and ISI (off time). This methodological power allows various aspects of attentional processing to be investigated, including:

- **Speed.** This can be determined from the search rate (i.e., the time required per item in the display). Note that the influence of salience can often be eliminated by the proper choice of items in the display, such as when half the items have one value of a property (e.g., vertical) and the other half the other value (e.g., horizontal).

- **Capacity.** This can be determined by increasing the amount of on time in each cycle. As on time increases, more items can be "grabbed" by attention, until saturation is reached. The value of this asymptote is a measure of attentional capacity (Rensink, 2000b). Note that this estimate is an upper bound on the number of attentional links involved because grouping factors may lead to chunking, causing more than one stimulus item to be assigned to each link.

- **Selectivity.** This can be determined by comparing the speed when all items must be examined against the speed when the change occurs in a selected subset. For example, the speed for orientation change can be measured for a set of black and white items; it can then be measured for the same items but with the change occurring only in the black ones. The ratio of these two speeds is a measure of the selectivity for black (Rensink, 1998).

- **Basic codes.** The basic "building blocks" (or codes) for coherent objects can be determined by comparing different types of changes and different kinds of items (see Rensink, 2000b). Changes in basic codes are indicated in three ways: (i) speed is relatively high because the least amount of coding and comparison is required; (ii) capacity is relatively high because only a minimal description needs to be stored; (iii) selectivity is efficient, since a minimum number of codes need to be excited or inhibited. It remains an interesting empirical matter to determine if the set of codes is equivalent to the set of "features" obtained from studies of search speed on static displays (e.g., Treisman and Gormican, 1988).

- **Task dependence.** Different tasks can be carried out on these flickering displays. These include not only detection (reporting if there is a change somewhere in the display) but also identification (reporting what the change is), and localization (reporting where it is). Although it might be thought that all these tasks should lead to similar estimates of attentional ability, such is not the case—for example, capacity estimates for identification are always below those for detection (Wilken et al., 1999).

In addition to this, the stimuli themselves can have different levels of complexity. Importantly, the level of complexity can be varied smoothly, allowing experiments to progress in a straightforward way from simple tasks on highly controlled arrays to more natural tasks on images of complex, real-world scenes.

9.2.2 Experimental results: Capacity

To illustrate how the approach described earlier can help map out the mechanisms of visual attention, consider the issue of attentional capacity (i.e., how many items can be attended at any one time). One way to determine this is by measuring the speed of detecting a changing target among non-changing distractors. If only one item can be attended at a time, search can only examine one item per alternation; the search rate then equals the alternation rate. If two items can be held, search should be twice as fast as the alternation rate. More generally, if h items can be held, search and alternation rates will be related by

$$\text{search rate} = (\text{alternation rate})/h \tag{9.1}$$

This can be rewritten as

$$h = (\text{alternation rate})/(\text{search rate}) \tag{9.2}$$

where the measure h (attentional hold) describes how many items on average are held and compared across the temporal gap. The attentional capacity C is the asymptotic value of h as display time increases.

Note that h is an average measure—it does not, for example, allow us to distinguish between h items processed at each alternation, $2h$ items every other alternation, or $3h$ items every third alternation. Only when h reaches its asymptotic value (i.e., the capacity) is it possible to state unequivocally that h items are being processed at each alternation.

In a series of experiments on search for changing orientation (Rensink, 1999, 2000b), search rates (and therefore values of h) were measured for a set of on times. The results (Fig. 9.6) show the existence of two different realms. The first is the set of on times of less than about 600 msec. Here, h increases linearly with on time, corresponding to a constant search rate (of about 100 msec/item); performance is evidently limited by the processing needed to read in and compare the items. As such, this realm is called the processing range. The second realm is the set of on times of more than 600 msec. Here, h is at an asymptote, indicating that performance is now governed primarily by the number of items that attention can hold. The particular value of the asymptote is about 5.5, indicating that information is collected from no more than 5 or 6 items at a time[8]

Consider now the reverse situation, where search is for a non-changing target among changing distractors (Rensink, 1999). As Figure 9.7 shows, two realms are again found: (i) for on times of less than about 300 msec, h increases linearly with ontime (corresponding to a rate of about 350 msec/item); (ii) for on times of more than 300 msec, h is at an asymptotic value of about 1.4. Although this estimate is

---

[8]Capacity measured in this way may include effects of grouping (or chunking). If used to estimate the number of stimulus items that can be held, this measure is perfectly suitable, being a simple operational description. However, if used to estimate the number of independent links involved, things are less straightforward because grouping will often cause more than one stimulus item to be assigned to a link. In this latter case, the measured capacity will be an upper limit on the number of links involved.

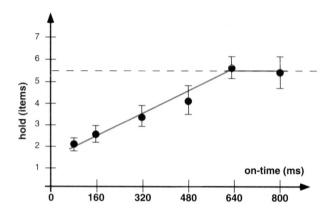

FIGURE 9.6. Attentional search for presence of change (orientation). Data from Rensink (2000b); off time for these conditions is 120 msec. Hold increases linearly with on-time up to about 600 msec, indicating that search rates are approximately constant in this realm. (Search speed is approximately 100 msec/item.) For longer on times, hold reaches an asymptotic value of 5.5 items.

above 1, it never reaches 2, suggesting that only one link at a time is being used. The excess of 0.4 items could result from a variety of factors, such as the grouping of neighboring structures.[9]

Evidence in support of this latter possibility is found in conjunction experiments. Here, the target changes in two dimensions (orientation and contrast polarity) simultaneously, whereas each distractor changes in just one (orientation or polarity). For this situation, the grouping of neighboring structures is less helpful. The pattern of two realms again emerges, as shown in Figure 9.8: (i) for on-times of less than about 300 msec, h increases with ontime (corresponding to a search rate of about 450 msec/item); (ii) for on-times of more than 300 msec, h again has an asymptote. However, the asymptote is now closer to 1 (Rensink, 1999), in accord with the reduced influence of grouping.

Bringing these results together, it appears that search for the presence of a change leads to a relatively high estimate of capacity (5-6 items), whereas search for the absence or conjunction of a change leads to a much lower estimate (1 item). Note that this pattern—obtained for attended structures—has an interesting similarity to that found for preattentive structures, where search for the presence of a distinctive feature is relatively easy, whereas search for the absence or the conjunction of a feature requires a slower, more effortful scan of the display (Treisman and Gormican, 1988).

---

[9]For example, consider the case where two adjacent items have the same orientation. It might be that these two items form a single group (pair) containing parallel items; if so, the pair could maintain its structure when the orientation of both its constituents changes. In this case, then, the detection of a nonchanging item could be signalled by a transition from a pair of parallel items to a pair of orthogonal ones.

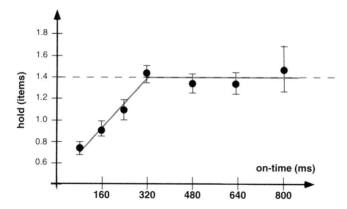

FIGURE 9.7. Attentional search for absence of change (orientation). Data from Rensink (1999); off time for these conditions is 120 msec. Hold increases linearly with increasing on time up to about 300 msec. (Search speed in this realm is approximately 300 msec/item.) The asymptote here is 1.4 items.

9.2.3 Implications for attentional mechanisms

The pattern of results for attentional capacity supports two proposals about the nature of focused attention: (i) the *singularity thesis*, the thesis that only one object is attended at any time (i.e., that there is only one nexus), and (ii) the *aggregation thesis*, the thesis that some of the information from each attended item is not kept separate (i.e., bound to that item), but is pooled into an aggregate description at the nexus.[10]

The justification for the singularity thesis rests on the finding that search for the absence of change has a severe capacity limit: 1.4 items. Although this value is slightly more than 1, it is clearly less than 2. Importantly, it stays constant over a fairly wide range of on times (300–800 msec). It may therefore be that this task involves only one link at a time, with the excess of 0.4 items due to factors such as grouping. It is worth pointing out that search in the processing range (i.e., search not limited by memory) takes about 300 msec/item; if grouping is taken into account, this corresponds to a "raw" rate of about $300 \times 1.4 = 420$ msec/item, a value close to the attentional dwell time (Ward, Duncan, and Shapiro, 1996). This suggests the involvement of a process that can handle only one item at a time,[11]

---

[10]Note that these theses are independent of each other. The singularity thesis concerns the number of "final destinations" for the pooled information, or equivalently, the number of mid-level representations that can be said to be genuinely independent. This could be one, but it could also be two, three, or some other positive integer. Meanwhile, the aggregation thesis concerns the degree and type of pooling done via the links. In principle, it could be, for example, that no pooling is done (i.e., only one link is used for each nexus) but that more than one nexus exists.

[11]It is possible to attend to more than one item in the image, at least if simple operations are involved (e.g., tracking the location of each item). However, the set of attended items feeds into the same nexus; it is not possible to treat them as completely separate objects. For example, when threading a needle both the needle and the thread must be attended. However, they are not independent objects, but parts

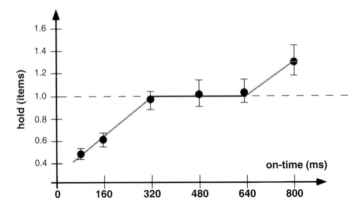

FIGURE 9.8. Attentional search for conjunction of change (orientation and contrast polarity). Data from Rensink (1999); off-time for these conditions is 120 ms. Hold increases linearly with increasing ontime up to about 300 ms. (Search speed in this realm is approximately 450 ms/item.) The asymptote here is close to just 1 item.

with the processing of the first item to be completed before the next can begin.

To see how the aggregation thesis can be justified, consider how information about attended items might be represented. One way is for each of the n links to maintain its own memory, with the link signalling to the nexus whenever a change to the corresponding proto-object occurs (*weak aggregation*); pooling is done only over these change signals (Figure 9.9). When searching for the presence of a change, the nexus signal will be 1 for target present and 0 for target absent, even when all links are pooled. When searching for the absence of change, however, the signal would be $n - 1$ for target present and n for target absent. If n is 2 or more, this would lead to a weaker signal. A more effective strategy would then be to monitor only one link at a time, which would yield a nexus difference of 1 vs. 0. Note that this argument is similar to that used to explain why in standard search the presence of a low-level feature is detected more quickly than its absence (Treisman and Gormican, 1988).

Another way that attended information might be represented is via the pooling of selected properties of the proto-objects themselves (*strong aggregation*), with the nexus containing an aggregate description, such as a distribution of pooled values (e.g., three of type "a" and two of type "b"; see Fig. 9.10). Search for the presence of change would involve detecting a change in this distribution, something that could be done if the comparison operation were sufficiently sensitive. (The limit on this sensitivity may be related to the limit on the number of links.) But search for the absence of change is far less tractable. For example, if one attended item is the target (i.e., constant) and the number of attended distractors is even the distribution

---

of a single needle-thread system. Note that this is not always counterproductive. In the case of the needle and thread, the fact that their locations are sent to the nexus allows accurate calculation of the relative distance between them, thereby facilitating the threading of the needle.

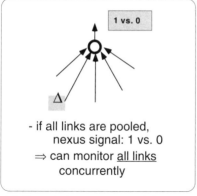
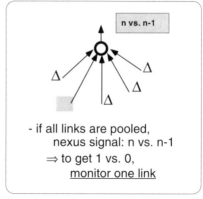

(a) Search for <u>presence</u> of change   (b) Search for <u>absence</u> of change

FIGURE 9.9. Explanation of search asymmetry (weak aggregation — pooling of change signals only). (a) When searching for the presence of change, the pooled signal will either be 1 (target present) or 0 (target absent). As such, there is a relatively strong signal at the nexus, even when information from several links is collected. (b) When searching for the absence of change, the pooled nexus signal will either be $n - 1$ (target present) or n (target absent). If n is more than 2 or 3, this signal would be quite weak. To obtain a strong signal, the nexus must collect information from only one link at a time.

may or may not change, depending on the particular items attended (Fig. 9.10). A better strategy might therefore be to monitor only one item at a time. In this case, the aggregate description is exactly that of the attended item and so can be directly used to determine whether or not that item is changing.

The finding that one item at a time is attended when detecting a conjunction of change leads to a further result: either the nexus gathers all the information from its links in parallel (thereby preventing the testing of individual links), or else each link fails to bind to it the different properties it contains. If neither of these were true—i.e., if all the properties of each link were bound to it, and if each attended link could be tested in turn—it would be possible to examine 5–6 items at each alternation, given sufficient time. The finding that only one item at a time is examined, however, means that at least one of these assumptions is incorrect. At the moment it is not known which of the two it is. It is worth noting, however, that the explanation based on lack of binding echoes the proposal for an absence of binding between the properties of each unattended low-level item, requiring an item-by-item search for a feature conjunction (Treisman and Gormican, 1988).

Although the results described earlier provide strong support for aggregation, they are not sufficient to determine if it is weak or strong. Settling this issue is an important task for future work. If aggregation is strong, it implies that focused attention acts by forming a partition between attended and unattended items, with the information from the attended items pooled into a single description. Note that such a scheme would solve the binding problem (i.e., the question of how to assign different properties to different attended structures) simply by having

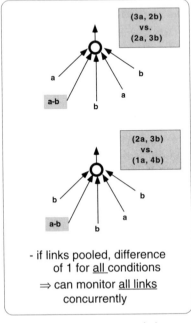 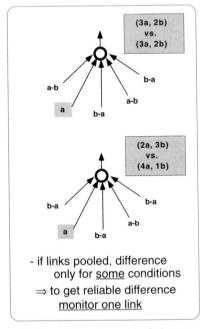

(a) Search for <u>presence</u> of change   (b) Search for <u>absence</u> of change

FIGURE 9.10. Explanation of search asymmetry (strong aggregation - pooling of proto-object properties). In these examples, a dash indicates change in successive displays; the letter to the left of the dash represents the first value, and the letter to the right the second. (a) Searching for the presence of change. For this situation, a change in one of the attended items always results in a change in the pooled signal (see upper and lower figures for a few examples). Thus, if the comparison process at the nexus is sensitive enough, a change in one of the links (i.e., that of the target) will always be detectable. (b) Searching for the absence of change. Here, aggregation will do one of several things, depending on the particular items linked. For example, for the values in the upper figure, the aggregate description stays the same; for the values in the lower figure, it changes considerably. Thus, monitoring the aggregate description will not always indicate when a change has occurred. To get a reliable signal of "no change" under all conditions, the nexus must collect information from only one link at a time.

no more than one discernible structure (i.e., the nexus description) in play at any time. Assigning properties to different parts of an object could be done by switching between individual parts as needed. If so, the perception of objects would occur via a virtual representation similar to that used for perceiving scenes.

9.3   Concluding Remarks

It has been argued here that the phenomenon of change blindness has considerable potential to cast light on the way that low-level descriptions link up with high-level knowledge to result in our perception of the world. This argument centered around the proposal that focused attention is needed to perceive change. Given this, it becomes possible to use change blindness to explore attention itself, both in regards to the role it plays in the perception of our surroundings, and to the mechanisms that underlie it.

In regards to the involvement of attention in scene perception, change-blindness studies show that we do not build up a detailed picture-like representation of the scene; rather, attention provides a coherent representation of only one object at a time. To account for the fact that we subjectively experience a large number of coherent objects simultaneously, it is suggested that scene perception involves a virtual representation which provides a limited amount of detailed, coherent structure whenever required, making it seem as if all the detailed, coherent structure is present simultaneously. In this view, vision is an inherently dynamic process – a "just in time" system whose detailed representations are in constant flux – and focused attention is an important part of its operation, providing coherent descriptions whenever requested.

It was also shown how the original flicker paradigm could be modified to explore various aspects of attentional processing. Initial results suggested that focused attention involves the collection of information into a single aggregate description. They also suggested that the information contained in this description is obtained via the pooling of information from a small number of links (no more than five or six). In addition, it appears that the properties obtained from the links may not be bound together, although further experiments are needed to establish this conclusively.

Taken together, these results lead to a marked shift in our understanding of mid-level vision. Rather than being a nebulous domain defined mostly in terms of what it is not,[12] it now becomes possible to map out several of its characteristics and to connect it with other systems known to be involved with scene perception. One of the main messages emerging from this new view is that mid-level vision is heterogeneous and highly dynamic. Consequently, there is little point in looking for a complete coherent representation of a scene or an object. Instead, it is likely

---

[12]The earlier characterization of mid-level vision tended to come perilously close to Gertrude Stein's famous quip about Oakland, California: "There is no there, there."

that only a small amount of information is ever put into coherent form at a time, with this amount being exactly that needed for the purpose at hand. (See also Ballard, 1991.)

In this view, then, mid-level vision is more concerned with the *management* of information than with its integration or storage. This raises issues such as how information is retrieved when requested, how much information is maintained in the representations that guide retrieval, and how operations can be coordinated to minimize collisions between responses to different requests. The solutions to these are, of course, unknown at the present time, but at least we have begun to ask these kinds of questions. Finding the answers may provide us with new insights into the mystery of how we turn the stream of photons entering our eyes into a vivid experience of our surroundings.

Acknowledgments

I would like to thank Ian Thornton and Carol Yin for their comments on an earlier draft of this chapter. I would also like to thank Carol Yin for her help in preparing QuickTime examples of the flicker paradigm for the accompanying CD-ROM. (Examples are also available at http://www.cbr.com/projects.)

References

Ballard, D. H. (1991). Animate vision. *Artif. Intell.*, 48:57-86.

Bridgeman, B., Hendry, D., and Stark, L. (1975). Failure to detect displacement of the visual world during saccadic eye movements. *Vis. Res.*, 15:719-722.

Grimes, J. (1996). On the failure to detect changes in scenes across saccades. In K. Akins (Ed.), *Perception* (Vancouver Studies in Cognitive Science, vol. 5, pp. 89-109). New York, NY: Oxford University Press.

Henderson, J. M. (1992). Object identification in context: The visual processing of natural scenes. *Can. J. Psych.*, 46, 319-341.

Henderson, J. M. and Hollingworth, A. (1999) High-level scene perception. *Ann. Rev. Psych.*, 50:243-271.

Kahneman, D., Treisman, A., and Gibbs, B. (1992). The reviewing of object files: Object-specific integration of information. *Cog. Psych.*, 24:175-219.

Klein, R., Kingstone, A., and Pontefract, A. (1992). Orienting of visual attention. In K. Rayner (Ed.), *Eye Movements and Visual Cognition: Scene Perception and Reading*, pp. 46-65. New York, NY: Springer.

Levin, D. T., and Simons, D. J. (1997). Failure to detect changes to attended objects in motion pictures. *Psychon. Bull. Rev.*, 4:501-506.

Marr, D. (1982). *Vision*. San Francisco: Freeman.

Moore, C. M, and Egeth, H. (1997). Perception without attention: Evidence of grouping under conditions of inattention. *J. Exp. Psych.: Human Percept. and Perform.*, 23, 339-352.

O'Regan, J. K, Deubel, H., Clark, J. J., and Rensink, R. A. (2000). Picture changes during blinks: Looking without seeing and seeing without looking. *Vis. Cog.*, 7:191-211.

Pashler, H. (1988). Familiarity and visual change detection. *Percept. and Psychophys.*, 44:369-378.

Pylyshyn, Z. W., and Storm, R. W. (1988). Tracking multiple independent targets: Evidence for a parallel tracking mechanism. *Spatial Vis.*, 3:179-197.

Rensink, R. A. (1998). Limits to attentional selection for orientation. *Percept.*, 27(suppl.):36.

Rensink, R. A. (1999). The magical number one, plus or minus zero. *Invest. Ophthalmol. Vis. Sci.*, 40:52.

Rensink, R. A. (2000a). The dynamic representation of scenes. *Vis. Cog.*, 7:17-42.

Rensink, R. A. (2000b). Visual search for change: A probe into the nature of attentional processing. *Vis. Cog.*, 7:345-376.

Rensink, R. A. (2000c). Seeing, sensing, and scrutinizing. *Vis. Res.*, 40:1469-1487.

Rensink, R. A., and Enns, J. T. (1995). Preemption effects in visual search: Evidence for low-level grouping. *Psychol. Rev.*, 102:101-130.

Rensink, R. A., and Enns, J. T. (1998). Early completion of occluded objects. *Vis. Res.*, 38:2489-2505.

Rensink, R. A. , O'Regan, J. K., and Clark, J. J. (1997). To see or not to see: The need for attention to perceive changes in scenes. *Psychol. Sci.*, 8:368-373.

Rensink, R. A. , O'Regan, J. K., and Clark, J. J. (2000). On the failure to detect changes in scenes across brief interruptions. *Vis. Cog.*, 7:127-145.

Simons, D. J., and Levin, D. T. (1997). Change blindness. *Trends Cog. Sci.*, 1:261-267.

Simons, D. J., and Levin, D. T. (1998). Failure to detect changes to people during a real-world interaction. *Psych. Bull. Rev.*, 5:644-649.

Stroud, J. M. (1955). The fine structure of psychological time. In H. Quastler (Ed.), *Information Theory in Psychology: Problems and Methods*. pp. 174-207. Glencoe, IL: Free Press.

Treisman, A., and Gormican, S. (1988). Feature analysis in early vision: Evidence from search asymmetries. *Psychol. Rev.*, 95:15-48.

Ullman, S. (1996). *High-level Vision*. Cambridge, MA: MIT Press.

Ward, R., Duncan, J., and Shapiro, K. (1996). The slow time-course of visual attention. *Cog. Psych.*, 30:79-109.

Wilken, P., Mattingley, J. B., Korb, K. B., Webster, W. R. and Conway, D. (1999). Capacity limits for detection versus reportability of change in visual scenes. Paper presented at the 26th Annual Australian Experimental Psychology Conference, April, 1999.

Wolfe, J. M. (1999). Inattentional amnesia. In V. Coltheart (Ed.), *Fleeting Memories*. pp. 71-94. Cambridge, MA: MIT Press.

Woodhouse, J. M., and Barlow, H. B. (1982). Spatial and temporal resolution and analysis. In H. B. Barlow and J. D. Mollon (Eds.), *The Senses*. pp. 133-164. Cambridge, England: Cambridge University Press.

10

The Role of Expectations in Change Detection and Attentional Capture

Daniel J. Simons
Stephen R. Mitroff

Imagine that you are driving your car and you come to a red light. While waiting, you play with the radio dial to find a more appealing station. You succeed, and return to waiting for the light to change. A few seconds later, the car behind you blares its horn. You look up only to realize that the light had changed to green while you were fiddling with the radio, and you did not notice when you looked up. Now imagine that you are trying to drive across a busy intersection. You carefully check for oncoming traffic in the near lane and then turn to check the far lane. Thinking your path is clear, you accelerate into traffic, only to realize that you had not seen the person in the crosswalk right in front of you. Take yet another case: you are a pilot trying to land an airplane under relatively foggy conditions. As you descend, you pay close attention to the elevation, speed, and pitch information projected onto your windshield's head-up display. You land the plane, never noticing there was another plane on the runway as you approached. Just this situation was studied in a flight simulator (Haines, 1991), and several professional pilots looked but did not see the other plane until they were too late to avoid it.

These examples have several things in common. First, all involve an attentionally demanding primary task (e.g., landing a plane, watching for traffic, etc.). Second, the unexpected object (e.g., the color of the stoplight, the pedestrian, or the plane) is not the currently attended aspect of the scene. Third, and most importantly for this chapter, the object change or appearance does not automatically capture attention when it is incidental to the primary focus of attention. Although we would like to believe that important, unanticipated events would automatically capture our attention, at least under these natural conditions, they sometimes do not.

Despite the importance of unexpected events for our understanding of attentional capture, research on attention generally has focused on the ability to detect (or ignore) briefly flashed objects when observers are intentionally looking for those objects. For example, the study of visual search has focused on how rapidly observers can spot an odd object among a set of different distractor objects (e.g., Treisman and Gelade, 1980). Similarly, the study of change detection has explored the ability to actively search for changes (e.g., Hollingworth and Henderson, 2000; O'Regan, Rensink, and Clark, 1999; Rensink, 2000; Rensink, O'Regan, and Clark,

1997; Simons, 1996; see Simons and Levin, 1997 for a review). Both lines of research have adopted what we will refer to as "the intentional approach." These
studies explore the psychophysics of attention and visual memory–under precisely controlled conditions, what are the limits on our capacity, speed, and precision
of detection? What are the thresholds of the attentional system? They determine
what we can remember from one view to the next or how much we can attend to
when we try. These traditional studies of attention and perception have focused
exclusively on intentional search or change detection. The primary task is to find
the target or the change. Yet, in the naturalistic examples discussed earlier, the "target" is secondary to the primary task, and detection is incidental to the observer's
primary goal.

The intentional approach has provided a number of important insights into the
limits of our abilities and the functioning of the visual system. However, such
tasks do not account for the entirety of visual experience. To describe human
performance fully, researchers need to adopt what we will refer to as "the incidental
stance." We need to know not just what the visual system can do, but what it actually
does. Although intentional visual search is central to our daily activities (e.g.,
finding your keys on a cluttered desk), the intentional approach may be suboptimal
for exploring the capture of attention by unexpected or unattended objects. By
adopting an incidental approach to the experimental study of attentional capture,
we may better understand how perception works in conditions such as landing
planes or driving through intersections.

Although the incidental approach has been relatively neglected in current empirical work on perception, recent research on change blindness from our lab (Simons
and Levin, 1997) and more recent literature on inattentional blindness (Mack and
Rock, 1998) have both adopted incidental approaches. Here we will review the
findings from these two bodies of literature in the context of intentional approaches to change detection and attentional capture, noting where the intentional and
incidental approaches produce similar results and where they differ. By combining
both approaches, we may gain a better understanding of the functioning and the
function of perception.

10.1   Change Blindness

Over the past few years, the study of change blindness has become a topic of
intense study in the field of visual cognition (see *Visual Cognition* 7(1/2/3) for a
special issue devoted to change blindness). In part, this focus on short-term visual
memory has resulted from a series of striking demonstrations of our inability to
notice large changes to complex displays (Grimes, 1996). The findings suggest that
we consciously perceive and remember far less of our visual world than we might
otherwise believe. The effects are particularly striking given our metacognitive
beliefs about perception and vision. We feel that we retain a rich representation of
our visual world and that large changes to our environment will draw our attention.

In fact, undergraduate students substantially overestimate their ability to detect change, predicting that they would notice changes that in fact few observers do (Levin, Momen, Drivdahl, and Simons, 2000). In this section, we briefly review evidence from the intentional approach to change detection, and then we discuss findings from the incidental approach.

10.1.1   The intentional approach

A comprehensive overview of the history of intentional approaches to change-detection is not central to the theme of this chapter (see Simons and Levin, 1997, for an overview of the literature on change blindness). Here we provide only a brief overview of more recent findings.

Intentional change detection has adopted two different paradigms. In *discrete* change detection tasks, observers view an initial display, followed by some disruption (e.g., an eye movement: Grimes, 1996; Henderson and Hollingworth, 1999; a blink: O'Regan, Deubel, Clark, and Rensink, 2000; or a blank screen: Simons, 1996; Rensink et al., 1997) and then by a modified version of the display. Subjects typically are required to report the location of the change or to determine whether or not a change occurred. This paradigm is similar to a traditional recognition-memory paradigm in that it allows for the use of signal-detection analyses and has a study/test format. It differs in that recognition memory tasks typically present a complete study set prior to the test but in these discrete change-detection tasks, the test display immediately follows the study display on each trial. Experiments using discrete change detection tasks have demonstrated that the ability to detect large changes to photographs or objects is surprisingly poor when the original and modified scenes are separated by a brief disruption. In all cases, observers are looking for changes but cannot easily find them if the change occurs during the interruption. Of course, when there is no disruption and the modified scene immediately follows the original image, observers readily detect the change. Without the disruption, the motion or luminance transient captures attention.

The second type of intentional task used to study change detection is the "flicker" paradigm (Rensink et al., 1997). An original and modified photograph alternate repeatedly, separated by a brief blank interval. The dependent measure is the number of cycles observers take to spot the change. Even with relatively large changes and after practice with the task, observers often take many cycles to notice the difference. In this paradigm, observers tend to notice changes to the central objects in a scene more than changes to peripheral objects even when the changes are physically equal in magnitude (Rensink et al., 1997). This paradigm has recently been adapted to study the capacity of visual memory and the speed of search (see Rensink's chapter in this volume and also Rensink, 2000). In many ways, the flicker paradigm is better than traditional search tasks in that observers know only the dimension of the change, but not the particular feature. For example, observers may search for a color change, rather than for a particular color. To the extent that researchers are interested in how efficiently and effectively observers can selectively attend to a feature dimension, searching for the dimension of a change is preferable to searching for a particular instantiation of that dimension.

Together, these intentional approaches have produced the following important results: (1) observers often miss large changes to simple displays and to photographs; (2) changes to central objects are noticed more quickly and more often than changes to peripheral objects; and (3) provided that the change coincides with some disruption (e.g., an eye movement, blank screen, eye blink, or blot), the transients that would otherwise capture attention are effectively masked. The fact that observers show change blindness when actively trying to spot changes suggests that they should be equally if not more blind to changes when change detection is not their primary task. However, intentional change detection requires subjects to encode scenes differently than they might under more natural encoding conditions. Given subjects' inaccurate metacognition about change detection (Levin et al., 2000), they might actually adopt less effective coding strategies when intentionally searching for change than when performing some other primary task.

10.1.2 The incidental approach

In the incidental approaches to change detection, observers are not forewarned that a change will occur and are simply asked to perform some other primary task. The displays in these tasks are often dynamic motion pictures or even real world events. A further difference from the intentional change detection tasks is that these incidental paradigms typically include only one critical test trial. Thus, the primary statistic is the proportion of people noticing a given change. This approach is not well-suited to the systematic study of the limits or thresholds of our change-detection abilities; systematic manipulation of the nature of the change in a one-trial study would not provide sufficient experimental power without a prohibitively large number of subjects. The primary purpose of these studies is to see whether the findings of intentional tasks generalize to real-world behaviors and to explore the extent of change blindness when we are not actively searching for change. These findings often provide striking examples of change blindness that appear to contradict our intuitive belief that we represent our world in great detail and that large changes will capture our attention. Further, they allow an exploration of the detection of changes to "attended" objects (Levin and Simons, 1997; Simons and Levin, 1997, 1998). Experiments using intentional change detection tasks consistently reveal better detection of changes to central objects in a scene (Rensink et al., 1997). Of course, in intentional tasks, observers are actively searching for change and are likely to focus on the central objects first. When they are searching for change, they encode the features of an object in order to try to detect something that differs. However, in the real world, we do not necessarily encode objects (even central objects) in such a way that would allow us to notice changes to features.

Although relatively few researchers have explored the perception of motion pictures, much attention has been paid to the fact that people fail to detect continuity errors or editing mistakes (Hochberg, 1986). People are often fascinated by these mistakes, and the popular media draws attention to them as if they were an anomaly.

For example, in a recent episode of ABC television's show *Dateline* that focused on editing errors in Academy Award-winning movies, the reporter asks "what is it about filmmakers that they can shoot so carefully, so many takes, and still miss something so obvious, something the audience can see clearly?" The reporter's intuition is that these changes are often noticed by the audience and that it is hard to imagine how the filmmaker failed to see them. Yet, in most motion pictures, continuity errors occur regularly despite the best efforts of the continuity editors. The anomaly is not that people miss these changes, but that they sometimes actually do see them. Levin and Simons have explored this aspect of change blindness empirically (Levin and Simons, 1997). In one experiment, they created a brief (approximately 1 minute) motion picture of a conversation between two actors. Every time the camera instantaneously changed (or "cut") to a new position, they intentionally inserted at least one editing mistake. For example, in one shot, a woman was wearing a scarf, but in the next shot it had disappeared, only to reappear in the following shot. Other changes included shifts to the positions of the actors' arms, a change in the color of plates on the table from red to white and then back to red, and a shift in the position of food from one plate to the other. They explored the ability to notice these changes by using an incidental task. Subjects were told to "watch this brief video and pay close attention." When the video ended, the subjects were asked if they had noticed any changes to the objects, clothing, or body positions in the video. Of the then subjects, only one reported noticing any of the nine intentional changes, and even that change was described only vaguely. Even on a second viewing, after being warned by the questions following the first viewing about what features to focus attention on, subjects still saw fewer than two of the nine changes on average. Consistent with findings from the intentional search tasks, subjects were generally unable to detect changes to a natural scene across a disruption (in this case, a film cut).

All the changes in this study were to peripheral objects, and we know from studies using the flicker task that peripheral changes are generally noticed more slowly than central changes. This incidental task can also be used to explore the detection of changes to central objects in a scene. In another version of the conversation film (Simons, 1996), the only change involved a central object. The scene opened with a shot of one actor pouring cola from a bottle into her cup. As she set down the bottle, the camera panned to show the approach of the other actor. When the camera panned back to the table, the bottle was gone and in its place was a cardboard box of roughly the same dimensions. When ten observers were asked to view this film and then to describe what they had seen, none noticed the change. Interestingly, several did mention the bottle in their descriptions of the movie. They just did not notice its disappearance. This finding suggests that even changes to the central object in a scene can go unnoticed if subjects are not explicitly looking for changes. The fact that observers described the bottle suggested that it was central to the scene and that they had focused attention on it during the film. However, the failure to describe the box suggested that it might not have been a central object when the camera returned. Perhaps subjects only notice changes when the target object is attended both before and after the change.

To investigate this possibility, another set of motion pictures was created in which the changed object was central throughout the film (Levin and Simons, 1997). A single character performed a simple action such as getting up to answer the phone or walking through a classroom and sitting in a chair. Given that the character was the only person in the film and was the only moving object, subjects are likely to attend to this character throughout the film (Dmytryk, 1984). During the action sequence, the camera cut to a new view of the character, and after the cut, the original actor was replaced with a different person who then completed the action. By cutting on the motion of the actor, the film appeared to depict a single, continuous event. Subjects were instructed to watch the brief (10sec) movie and to pay close attention. Immediately after the movie ended, subjects wrote a description of what they had seen. Any indication of change detection in these written descriptions was taken to indicate successful detection. In other words, subjects did not need to report that the person had changed identity, but only to note that something about the actor had changed (there were also clothing differences between the two actors). Across eight different change films (with four pairs of actors), on average only 35% of subjects noticed a change. The written responses often included detailed descriptions of the actors and always mentioned the person both before and after the cut, suggesting the actor was the central object throughout the film. Yet, most subjects failed to see the change. Under incidental encoding conditions, observers fail to represent or compare those features that change from one view to the next. Another set of observers viewed all of the change films and an equal number of no-change films under intentional conditions – they successfully detected which films included a change.

Under incidental conditions, focusing attention on the central object does not guarantee that all of the features of that object are encoded and retained (see Simons and Levin, 1997). This result appears discrepant from a recent finding that used an intentional task with simple arrays of objects (Luck and Vogel, 1997). In this short-term memory task, observers tried to remember the properties of up to four objects. Although they generally could only remember four individual features in isolation, they could remember many more if the features were tied to individual objects. In other words, subjects could remember all of the properties of each of four objects; performance was limited by the number of objects and not the number of features in the display, and attention to an object allowed subjects to remember all of its feature dimensions. This finding raises an interesting possibility: perhaps observers only retain those properties that are the focus of their intentional efforts to remember. When subjects know they could be tested on any of a number of feature dimensions, they encode the display so that they will remember those features more accurately. In contrast, when they do not know that they may be tested, they encode only those features necessary to comprehend the meaning. They do not automatically encode those features that would discriminate one actor from another.

Although motion pictures are a good medium for studying incidental change detection and they are closer to actual experience than simple arrays or photographs, they still may not adequately test how we encode objects in the real world. When encountering another person, we may well encode and retain more information than we would from a brief glimpse of an actor in a passively observed motion picture. Simons and Levin tested this possibility by using a real-world change detection task (Simons and Levin, 1998). In these studies, an experimenter approached a pedestrian (the subject) to ask for directions. During their conversation, two people carrying a door passed between the experimenter and the pedestrian, and during that interruption, the first experimenter was replaced by a second experimenter who was wearing different clothing. Even though subjects engaged in an interaction with both the first and the second experimenter, 50% of the subjects did not notice the person change. The two experimenters in this study were both approximately 30-year-old white men with short dark hair, but in other respects they were quite different. For example, they were 6 cm different in height, and the shorter experimenter had a much deeper voice.

Interestingly, the 50% who missed the change (typically faculty or staff) were all older than the experimenters and the 50% who noticed (typically students) were all the same age or slightly younger than the experimenters (Simons and Levin, 1998). Although this effect could be attributed to aging, most of the "older" subjects were not elderly. It seemed more likely that the effect resulted from a difference in how people encode members of their own social group as opposed to members of a different social group. When viewing members of their own social group, people tend to encode features that differentiate individuals. In contrast, when viewing members of a different social group, people tend to encode group membership information and to ignore differentiating features (Rothbart and John, 1985). The older subjects simply coded the experimenter(s) as a student asking directions. Hence, they had not encoded any features that would change as a result of the switch in experimenters. In contrast, the younger subjects were more likely to encode features that would differentiate individuals, so they noticed the different features when the experimenters changed. To explore this possibility further, the experimenters were made members of an outgroup to the younger subjects by dressing them as construction workers. In this condition, only 35% of a new group of younger subjects noticed the change, down from 100% in the earlier version (Simons and Levin, 1998). This ingroup/outgroup difference likely would not have been noticed using an intentional change detection task. If observers knew that the person might change, they would likely code individuating features regardless of their group membership. When observers are not actively searching for a change, they tend to focus on the meaning of a scene – what is important for their immediate actions and goals. Such encoding is unlikely to lead subjects to focus on specific visual details. As a result, they are even less likely to detect changes, even if the changes are to central objects.

Although these findings could be taken to suggest that relatively little if any visual information is retained in daily experience, this conclusion would be premature and based on faulty logic. Successful change detection across an interruption does require a representation of both the initial and changed objects (or at least of the difference between them). However, change blindness does not imply the absence of a representation (see Simons, 2000a for a discussion). Subjects could fail to detect a change because they lack a representation of what had been present before, or they may be change-blind because they accurately represented the initial scene but not the modified scene. They may even accurately represent the features of both the initial and changed scene but still not detect the change if they fail to compare the scenes in the appropriate way (Scott-Brown, Baker, and Orbach, 2000; Simons, 2000a). In the real world case, subjects may not have made such a comparison because the change did not produce any inconsistency with the meaning of the scene and it did not capture attention. Further, the visual system likely assumes stability, so unless a transient signals a change, there would be no reason to try to compare the initial and modified representations. Several suggestions from research with motion pictures are consistent with the possibility that we actually do represent the details of the changed object when change detection fails. For example, in the bottle-to-box change (Simons, 1996), observers often described the bottle after the film ended. If change blindness results from the overwriting of the initial object by the changed object, subjects should not have been able to describe the bottle. Similarly, a number of subjects in the person-change films described the properties of the first actor rather than the second actor (Levin and Simons, 1997).

A more recent real-world experiment also examined the nature of our representations when change detection fails (Simons, Chabris, and Levin, 1999). In this study, an experimenter approached a pedestrian (the subject) to ask directions to a gymnasium. The experimenter was holding a red and white striped basketball and was wearing gym clothes. As the pedestrian was providing directions, a group of confederates walking on the sidewalk passed between the experimenter and pedestrian, and one of them surreptitiously took the basketball. If the initial representation is replaced following the disruption, subjects who fail to notice the change should not remember the basketball. Based on a series of probing questions, only 20% of subjects spontaneously reported the change. Surprisingly, when asked directly whether they thought the experimenter originally had a basketball, more than 50% said yes. Furthermore, most were able to describe the unusual appearance of the ball and none described a canonical basketball. Results of a no-change control condition suggest that this finding does not result from the leading questions. In essence, people may have accurate representations of both the original and modified features and still fail to notice changes.

10.1.3 Summary

Both incidental and intentional approaches have produced striking evidence for change blindness. Observers often fail to detect large changes to photographs or scenes from one view to the next, even when they are actively looking for changes. Although both intentional and incidental approaches provide evidence for change blindness, neither approach alone completely captures the phenomenon. Evidence from the intentional approach suggests that changes to central objects are detected more readily than changes to peripheral objects. From this finding, we might be tempted to conclude that change detection is driven solely by how central an object is to the scene. We would be tempted to infer that attention is drawn to central objects and that attention is sufficient for the accurate representation and detection of changes. Yet, evidence from the incidental approach shows that even changes to attended objects can go undetected, suggesting that attention to an object is not sufficient for change detection. If we only considered evidence from the incidental approach, we would be similarly misled. Although we might intuitively expect some benefit of focused attention, the incidental approach is ill-suited for systematic exploration of the effects of centrality in a scene. If we relied only on findings from the incidental approach, we might incorrectly conclude that changes to central objects are no more likely to be detected than other changes. By adopting the intentional approach, we can gain a better understanding of our change-detection mechanisms, and by adopting the incidental approach, we can gain a better appreciation for how those mechanisms operate in the real world. By combining both approaches, we can avoid the pitfalls engendered by adopting either approach in isolation.

10.2   Attentional Capture

Perhaps the only domain to be studied exclusively using incidental tasks is that of "inattentional blindness." Inattentional blindness refers to the tendency not to see unattended objects (Mack and Rock, 1998). The study of inattentional blindness has its roots in research on selective attention and dichotic listening (Moray, 1959; Treisman, 1960, 1964; Neisser and Becklen, 1975; Neisser, 1979; Becklen and Cervone, 1983; Holender, 1986; Stoffregen and Becklen, 1989). Such studies focus on the ability to report what was presumably unattended information, often under conditions in which subjects do not expect any information to be present. Similarly, most studies of inattentional blindness explore the perception of unattended and unexpected information (e.g., Mack and Rock, 1998; Newby and Rock, 1998; Rock, Linnett, Grant, and Mack, 1992). In essence, these studies explore the tendency for different forms of information to capture attention when subjects have no prior expectation that the information will be presented at all. Once subjects become aware of the unexpected object, event, or sound (either because they noticed it or because they were questioned about it), they now expect the unexpected and the nature of the task has changed fundamentally. Consequentially, most studies of inattentional blindness have relatively few critical test trials. Because intentional tasks are rarely useful when they have only one trial, inattentional blindness has been explored almost exclusively with incidental tasks. However, the empirical study of attentional capture has employed intentional tasks, perhaps the most prominent of which is visual search (see Simons, 2000b for a more complete review of the links between attentional capture and inattentional blindness).

10.2.1 The intentional approach

Intuitively, it would seem sensible for a visual system to detect unexpected or unusual events automatically, drawing attention to them, thereby allowing a rapid and appropriate response. For example, if a predator suddenly charged toward you, you would want to become aware of it without having to consciously and effortfully shift attention to it. Because we cannot introduce predators into a laboratory setting, researchers have adopted the methods of visual search as a proxy for this more natural situation. Visual-search studies of attentional capture look for indications that a target object was detected without any effort or that it drew attention away from other items (Yantis and Jonides, 1984; Folk, Remington, and Johnston, 1992; Theeuwes, 1994; Yantis, 1998). Effort in a search task is operationally defined by the effect of this target item on search latencies. If it captures attention, it should be processed before other items in the display. Thus, if the distinctive item is the target of the search task, then search latency will be unaffected by the number of other items in the display. Similarly, if the distinctive item is one of the distractor items, search latency will be relatively slowed because attention will be drawn to this incorrect item rather than to the target of the search.

Several different variants of this search task have been developed. In the "irrelevant-feature search task," observers perform a traditional search task (e.g., looking for an L among Ts), and one item in the display is different from all of the other items (e.g., it onsets later or is a different color). This distinctive item does not predict the location of the search target; it is no more likely to appear as the target than as one of the distractors. Observers know that the distinctive feature is irrelevant and that there is no reason to focus attention on it – it will not aid or impair their search performance. If this irrelevant feature captures attention when it happens to coincide with the search target, detection latency should be unaffected by the number of other items in the display. This methodology allows a systematic exploration of the sorts of features that might capture attention. Across a number of studies, the one feature that consistently appears to capture attention is a sudden and late onset of the target item (see Yantis and Hillstrom, 1994; Yantis and Jonides, 1996, although there is still debate about whether this result is based entirely on bottom-up attentional capture or whether the observer's attentional set influences search and contributes to speeded search; see, Folk et al., 1992; Yantis, 1993). Abrupt onsets reduce search latencies, suggesting that they can automatically draw attention.

In another intentional approach to attentional capture, observers view an irrelevant spatial cue prior to a search task. The cue does not predict the target location. Yet, when the properties of this cue match those of the target (e.g., a color precue for a color target), and if it happens to appear in the location of a target, observers show speeded search performance (Folk, Remington, and Wright, 1994). Similarly, when the spatial precue happens to appear in the location of a distractor item, performance is impaired. Presumably, attention is drawn to this cue automatically. When the cue is consistent with the nature of the search (e.g., subjects are looking for a color target and they receive a color cue), they cannot help but pay attention to it even though it is known to be irrelevant to the primary task (Folk et al., 1994).

In both of the preceding cases, observers perform an intentional search task, fully expecting this irrelevant item to appear on every trial. They know that focusing attention on the distinctive feature will not improve their performance. Yet, they still shift attention to it. Such findings have been taken to suggest that some distinctive features (e.g., abrupt onsets) draw attention automatically. Even though the feature is irrelevant, observers cannot help but attend to it.

Unfortunately, an intentional task may not adequately reflect the sorts of attentional capture that we would need in order to avoid a charging predator. These intentional tasks effectively focus on the ability not to attend to something that we know to be irrelevant. The distinctive item is always present but does not help in the primary task performance. In the real world, however, the predator does not always appear, and when it does, it is rarely irrelevant or expected. Do such unexpected stimuli capture attention? Only one experiment using the irrelevant-feature search paradigm has explored this question (Gibson and Jiang, 1998). For the first 192 trials of this study, observers performed a traditional conjunction search task. On the 193rd trial, one of the items in the search display was a different color from the others. Yet, this distinctive and unexpected item failed to capture attention; search performance was no different from what was predicted by the preceding trials. The absence of attentional capture in this study raises a somewhat radical possibility: in the absence of expectations, unusual and distinctive objects may not capture attention. In other words, attention may not automatically be drawn to the sudden appearance of a predator. The intentional approach is not well-suited to explore whether unexpected objects capture attention. Hence, recent experiments on inattention have adopted incidental approaches.

10.2.2   The incidental approach

Studies of inattentional blindness have used two different paradigms. In an approach introduced by Mack and Rock (Mack, Tang, Tuma, and Kahn, 1992; Rock et al., 1992; Mack and Rock, 1998), subjects engaged in a primary task of determining which line of a cross (the horizontal or vertical) was longer. On each trial, the cross was flashed for 200 msec. On the critical trial (typically the third or fourth trial), another object appeared along with the cross. Subjects were then asked to report whether they had seen the unexpected object. Attentional capture is indicated by higher rates of detection, and inattentional blindness is indicated by failed detection. Once subjects were asked about the unexpected object, they knew to look for it on subsequent trials. A subsequent trial with the object was therefore a divided-attention trial (divided between searching for anything other than the cross and performing the line-judgment task). Subjects were consistently better able to detect the object on these divided attention trials, suggesting that the absence of expectations plays a role in inattentional blindness.

Experiments using this paradigm reveal a surprising degree of blindness for the unexpected object (Mack and Rock, 1998). Even when the object is a different color it is no more likely to be noticed than a black object (both are noticed by 25 – 75% of subjects, depending on the conditions). Of course, there are differences in which features are noticed and which are not. First, objects that are relatively close to the focus of attention (i.e., the cross) are more likely to be noticed (Mack and Rock, 1998). Furthermore, stimuli that are meaningful to observers are often noticed (Newby and Rock, 1998). For example, subjects are more likely to notice their own name than they are their own name with one letter changed (e.g., Jake vs. Jeke). They also tend to notice some schematic objects such as smiley faces.

One concern about drawing strong conclusions from Mack and Rock's line-judgment task is that the target object is flashed only briefly. Under more natural conditions, unexpected objects would likely be visible for more than 200 msec, and subjects would not be fixating a single point. For example, pilots in the flight-simulator study described at the beginning of the chapter had an extended opportunity to view the unexpected target object (Haines, 1991).

Studies using a different paradigm, "selective looking," have explored the capture of attention by unexpected objects and events in displays that last substantially longer than 200 msec (Neisser and Becklen, 1975; Neisser, 1979). In these studies, observers view a display with two ongoing events, but they are only required to monitor one of them. During their ongoing performance of this task, an unexpected event occurs. At the end of the trial, subjects are asked to report what they saw.

This method for studying attentional capture was developed by Ulric Neisser and colleagues during the 1970s and 1980s (Neisser and Becklen, 1975; Littman and Becklen, 1976; Neisser, 1976; Neisser, 1979; Becklen and Cervone, 1983; Stoffregen and Becklen, 1989). For example, Neisser and Becklen (1975) showed observers two ongoing events, each partially transparent and superimposed on top of the other. Thus, the events occupied the same spatial locations on the display and therefore on the retina. In this initial experiment, one event was a hand slapping game with two players and the other was a group of three people passing a basketball to each other while moving around. Subjects were required either to press a button whenever the basketball players made a pass or to press a button when a particular event happened in the hand game (in some cases, subjects tried to do both tasks simultaneously). After several trials of this task, something unexpected occurred in the unattended event. For example, the people playing the hand game would stop and shake hands or the people in the basketball game would temporarily have no basketball but would continue to fake passes. Surprisingly, many subjects failed to notice these unexpected events (see Neisser, 1979).

In a further variant of this experiment (Becklen and Cervone, 1983), the two superimposed events were identical basketball games with one group of players wearing white and the other wearing black. As subjects were selectively monitoring one team, a woman carrying an open umbrella walked slowly across the display and off the far side. All three events were partially transparent, and because the umbrella woman walked across the screen, she occupied the same spatial locations as the players. Yet, even though this unexpected event lasted nearly 5 sec, many subjects failed to notice it.

Neisser and colleagues conducted a number of additional variants on this task, but unfortunately, many of them were not published and were described only in a summary book chapter (see Neisser, 1979). The original work on these issues was difficult to incorporate into the theory of the time (see Simons, 2000), but in light of more recent work on inattentional blindness it has taken on new significance (Simons, 1999; 2000b). Not only can we miss briefly flashed objects away from the focus of attention, but we also fail to see ongoing dynamic events, provided that attention is focused on another event or object. Furthermore, attention in this task need not be focused on a different spatial location to produce inattentional blindness, an important difference from work using the line-judgment task (Newby and Rock, 1998). In selective looking, attention seems to be directed to objects and events rather than to spatial locations.

Interestingly, in both the line-judgment and selective looking tasks, stimuli that might capture attention in an intentional task do not do so (Simons, 2000b). This fundamental difference has important implications for models of visual capture and visual representations. Color singletons may only capture attention if we know they may appear. Models of attention and vision are often based on the notion that some features will pre-attentively capture attention (Treisman and Gelade, 1980). Yet, if under more typical viewing conditions, features thought to be "primitive" fail to capture attention, these models will only account for performance when subjects are aware that a stimulus may appear. The idea behind intentional studies of attentional capture is that unexpected or unusual objects should capture our attention. Yet, the methods used to study capture may be sub-optimal for testing whether they do capture attention.

Both the line-judgment and selective looking tasks suggest that without attention, some objects and events will not be consciously perceived (see Mack and Rock, 1998). However, neither speaks particularly well to the perception of u-nattended objects in naturalistic displays. The line-judgment task involves stimuli flashed only briefly on a computer monitor and typically does not involve a dynamic unexpected event (with the exception of briefly presented stroboscopic motion sequences, see Mack and Rock, 1998). The superimposed selective looking task is also somewhat unnatural. Although subjects could see the umbrella woman if they looked for her (Becklen and Cervone, 1983), perhaps the limited detection is due to the unnatural transparent display. Perhaps inattentional blindness only occurs for briefly flashed displays or for events that are difficult to view.

Simons and Chabris in the late 1990's conducted a new series of studies in an effort to revive interest in the study of inattentional blindness for prolonged, dynamic events as a tool for understanding the perception of unattended objects (Simons and Chabris, 1999). These studies used a task quite similar to those used by Neisser and colleagues. Observers viewed a dynamic display of one team of three players in white shirts and one team in black shirts, each passing a basketball. Subjects counted how many passes one of the teams made. In the easy version of the task, they simply counted the total number of passes. In the difficult version, they kept two running totals, one of the number of aerial passes and one of the number of bounce passes. After approximately 45 seconds of performing this task, an unexpected event occurred. Two distinct unexpected events were used: a woman carrying an open umbrella (as in the original studies by Becklen and Cervone, 1983) and a woman wearing a black gorilla suit. Finally, to explore the possibility that the degree of inattentional blindness in the earlier dynamic displays resulted from the odd superimposition, two versions of each event were used. In one version, both teams and the unexpected event were superimposed and partially-transparent, much as they had been in the earlier studies. In the other version, both teams and the unexpected event were filmed using a single camera and choreographed action. In this ³opaque² version, all of the objects were fully visible and clear, and the players could occlude one another. Consistent with the earlier work using transparent dynamic displays, observers often did not see the partially-transparent umbrella woman or the gorilla when their attention was focused on another object or event. Averaging across both unexpected events, both colors of attended team, and both easy and difficult tasks, nearly 60% missed the unexpected event when the teams were partially transparent, a level roughly consistent with the findings of Neisser and colleagues (Neisser, 1979). Yet, even when the teams and unexpected object were fully visible and opaque, nearly 35% still did not see them! As expected, more subjects noticed the unexpected object when performing the easy counting task than the difficult counting task.

One interesting and unexpected finding was that more people noticed the gorilla when counting the passes made by the team wearing black than when counting passes by the team wearing white. This difference may be due to the greater similarity of the gorilla to the black team than the white team (the umbrella woman was wearing brown clothes so she was different from both the attended and ignored teams). To explore this possibility, our lab recently conducted a new series of studies using more tightly controlled computer displays (Most et al., accepted). In these experiments, subjects viewed a display with four white shapes and four black shapes. The shapes moved pseudo-randomly, periodically bouncing off the sides of the display window. The subject's task was to count the total number of times that one set of shapes (either black or white) bounced off the sides. Each trial of this task lasted for 15 sec, and on the third trial, an unexpected event occurred: after 5 sec, a cross began to move linearly across the center of the display, passing the fixation point, and exiting the other side of the display 5 sec later. After this trial, subjects were asked whether they had seen the unexpected object. Unlike the video experiments, in this study the similarity of the unexpected object to the attended and ignored items could be controlled precisely by varying the luminance of the cross.

The results bore out the effect of similarity we found in the video studies: when attending to the white shapes, subjects generally noticed a white cross and almost never noticed a black cross. Similarly, when attending to black shapes, they almost always noticed a black cross and almost never noticed a white cross. Detection of gray unexpected crosses was intermediate and noticing varied with the similarity of the luminance to the attended items and the dissimilarity to ignored items.

These findings clearly show the importance of visual similarity in the detection of unexpected objects. However, based on this effect of luminance alone, we cannot determine whether detection is based on how similar the cross is to the attended items or how different it is from the ignored items – the two explanations were perfectly confounded. In a final experiment, subjects attended to a set of gray shapes against a colored background and ignored either a set of white shapes or a set of black shapes (only one ignored set was present in a given experiment). In this case, the unexpected cross was either white or black; it was either the same as the ignored items or it was equally different from the attended objects but in a direction different from the ignored items. Surprisingly, subjects were less likely to notice the cross when it was the same color as the ignored items than when it was different from the ignored items. In these cases, the unexpected object was equally different from the attended items. This finding suggests an important role for the ignored items in determining what captures attention. Subjects apparently are actively inhibiting conscious perception of items similar to the ignored items when performing a selective attention task (Watson and Humphreys, 1997, 1998). It is interesting to note that subjects were not required to ignore the irrelevant items. They were just told to selectively attend to one group of items. Our lab is currently exploring whether inattentional blindness for dynamic events requires that subjects selectively ignore some of the items or if they will still fail to notice unexpected objects, provided that attention is focused on other objects or events.

10.2.3   Implications

Subjects often fail to see unexpected objects in situations ranging from brief flashes on a computer display to ongoing, naturalistic, dynamic events. This failure seems to depend on two factors: (1) observers do not expect the object or event, and (2) observers are engaged in an attentionally demanding task. In the line-judgment task, observers do not selectively ignore any other items, but their attention is focused on the cross. Under these conditions, inattentional blindness is related to the proximity to the cross. In the dynamic tasks, subjects selectively ignore one set of items while attending to other items. Under those conditions, we find inattentional blindness even when the unattended and attended items occupy the same space. However, the similarity in appearance of the unexpected object to the attended and ignored items does seem to play a role.

Although these two paradigms produce somewhat different results, the overall pattern using these incidental tasks is consistent and, more importantly, different from the results of intentional tasks. When observers do not expect an object to appear, they often do not see it at all. Yet, in an intentional search task, observers often see the odd object – it captures attention. The factors that produce attentional capture in an intentional task do not necessarily produce the same results in an incidental task. The fundamental difference appears to be the expectations of the observer. In the real world, our attention is typically focused on some goal and we do not always expect the unexpected.

10.3   Conclusions

The *incidental stance* approach to perception asks what people do under typical perceptual conditions. Models based solely on intentional tasks may describe the capabilities and capacities of attention, but they may not adequately describe what people actually perceive and how they perceive it. For example, models of perception that require the capture of attention by primitive features may not generalize to naturalistic viewing conditions; attentional capture may depend on the observer's expectation that something might suddenly appear. Similarly, when observers do not intentionally search for changes, they often fail to notice when the central object in a scene is replaced. If we relied exclusively on intentional change-detection tasks, we might have concluded that changes to central objects are typically detected.

Of course, without intentional tasks, our understanding of attention would be quite limited. Incidental tasks are ill-suited to the exploration of the systematic variation in performance and cannot provide much information about sensory or attentional thresholds. With only one critical trial, the incidental approach is fundamentally limited to studies of the average performance of groups of subjects rather than the performance of individual subjects. By using both the intentional and incidental approaches in concert, we can gain a better understanding of the full range of change blindness and of visual perception in general.

Acknowledgments

See the enclosed CD rom for videos from some of the incidental tasks described in this chapter. Videos are also available on our laboratory web site http://www.mjh.harvard.edu/~ viscog/lab. The writing of this manuscript was supported in part by NSF grant #BCS-9905578 and by a Research Fellowship from the Alfred P. Sloan Foundation.

References

Becklen, R., and Cervone, D. (1983). Selective looking and the noticing of unexpected events. *Mem. Cognit.*, 11:601-608.

Dmytryk, E. (1984). *On Film Editing: An Introduction to the Art of Film Construction.* Boston, MA: Focal Press.

Folk, C. L., Remington, R. W., and Johnston, J. C. (1992). Involuntary covert orienting is contingent on attentional control settings. *J. Exp. Psych. Hum. Percept. Perf.*, 18:1030-1044.

Folk, C. L., Remington, R. W., and Wright, J. H. (1994). The structure of attentional control: Contingent attentional capture by apparent motion, abrupt onset, and color. *J. Exp. Psych. Hum. Percept. Perf.*, 20:317-329.

Gibson, B. S., and Jiang, Y. (1998). Surprise! An unexpected color singleton does not capture attention in visual search. *Psychol. Sci.*, 9:176-182.

Grimes, J. (1996). On the failure to detect changes in scenes across saccades. In K. Akins (Ed.), *Perception, Vancouver Studies in Cognitive Science* Vol. 2, pp. 89-110. New York: Oxford University Press.

Haines, R. F. (1991). A breakdown in simultaneous information processing. In G. Obrecht and L. W. Stark (Eds.), *Presbyopia Research: From Molecular Biology to Visual Adaptation*, pp. 171-175. New York, NY: Plenum.

Henderson, J. M., and Hollingworth, A. (1999). The role of fixation position in detecting scene changes across saccades. *Psychol. Sci.*, 10:438-443.

Hochberg, J. (1986). Representation of motion and space in video and cinematic displays. In K. R. Boff, L. Kaufman, and J. P. Thomas (Eds.), *Handbook of Perception and Human Performance Vol. 1: Sensory Processes and Perception*, pp. 22.21-22.64. New York, NY: John Wiley and Sons.

Holender, D. (1986). Semantic activation without conscious identification in dichotic listening, parafoveal vision, and visual masking: A survey and appraisal. *Behav. and Brain Sci.*, 9:1-66.

Hollingworth, A., and Henderson, J. M. (2000). Semantic informativeness mediates the detection of changes in natural scenes. *Vis. Cog.*, 7:213-235.

Levin, D. T., Momen, N., Drivdahl, S. B., and Simons, D. J. (2000). Change blindness blindness: The metacognitive error of overestimating change-detection ability. *Vis. Cog.*, 7:397-412.

Levin, D. T., and Simons, D. J. (1997). Failure to detect changes to attended objects in motion pictures. *Psychon. Bull. Rev.*, 4:501-506.

Littman, D., and Becklen, R. (1976). Selective looking with minimal eye movements. *Percept. Psychophys.*, 20:77-79.

Luck, S. J., and Vogel, E. K. (1997). The capacity of visual working memory for features and conjunctions. *Nature*, 390:279-281.

Mack, A., and Rock, I. (1998). *Inattentional blindness*. Cambridge, MA: MIT Press.

Mack, A., Tang, B., Tuma, R., and Kahn, S. (1992). Perceptual organization and attention. *Cog. Psych.*, 24:475-501.

Moray, N. (1959). Attention in dichotic listening: Affective cues and the influence of instructions. *Q. J. Exp. Psych.*, 11:56-60.

Most, S. B., Simons, D. J., Scholl, B. J., Jimenez, R., Clifford, E., and Chabris, C. F. (accepted). How not to be seen: The contribution of similarity and selective ignoring to sustained inattentional blindness. *Psychol. Sci.*

Neisser, U. (1976). *Cognition and reality: Principles and Implications of Cognitive Psychology*. San Francisco, CA: W. H. Freeman.

Neisser, U. (1979). The control of information pickup in selective looking. In A. D. Pick (Ed.), *Perception and its development: A tribute to Eleanor J. Gibson*, 201-219. Hillsdale, NJ: Lawrence Erlbaum.

Neisser, U., and Becklen, R. (1975). Selective looking: Attending to visually specified events. *Cog. Psych.*, 7:480-494.

Newby, E. A., and Rock, I. (1998). Inattentional blindness as a function of proximity to the focus of attention. *Percept.*, 27:1025-1040.

O'Regan, J. K., Deubel, H., Clark, J. J., and Rensink, R. A. (2000). Picture changes during blinks: Looking without seeing and seeing without looking. *Vis. Cog.* 7:191-211.

O'Regan, J. K., Rensink, R. A., and Clark, J. J. (1999). Change-blindness as a result of "mudsplashes". *Nature*, 398:34.

Rensink, R. A. (2000). Visual search for change: A probe into the nature of attentional processing. *Vis. Cog.* 7:345-376.

Rensink, R. A., O'Regan, J. K., and Clark, J. J. (1997). To see or not to see: The need for attention to perceive changes in scenes. *Psychol. Sci.*, 8:368-373.

Rock, I., Linnett, C. M., Grant, P., and Mack, A. (1992). Perception without attention: Results of a new method. *Cog. Psych.*, 24:502-534.

Rothbart, M., and John, O. P. (1985). Social categorization and behavioral episodes: A cognitive analysis of the effects of intergroup contact. *J. Soc. Iss.*, 41:81-104.

Scott-Brown, K. C., Baker, M. R., and Orbach, H. S. (2000). Comparison blindness. *Vis. Cog.* 7:253-267.

Simons, D. J. (1996). In sight, out of mind: When object representations fail. *Psychol. Sci.*, 7:301-305.

Simons, D. J. (1999). To see but not to see: Review of Inattentional Blindness by A. Mack and I. Rock (1998). *J. Math. Psych.*, 43:165-171.

Simons, D. J. (2000a). Current approaches to change blindness. *Vis. Cog.* 7:1-15.

Simons, D. J. (2000b). Inattentional blindness and attentional capture. *Trends Cog. Sci.*.

Simons, D. J., and Chabris, C. F. (1999). Gorillas in our midst: Sustained inattentional blindness for dynamic events. *Percept.*, 28:1059-1074.

Simons, D. J., Chabris, C. F., and Levin, D. T. (1999). Change blindness is not caused by later events overwriting earlier ones in visual short term memory. manuscript in preparation.

Simons, D. J., and Levin, D. T. (1997). Change blindness. *Trends Cog. Sci.*, 1:261-267.

Simons, D. J., and Levin, D. T. (1998). Failure to detect changes to people in a real-world interaction. *Psychon. Bull. Rev.*, 5:644-649.

Stoffregen, T. A., and Becklen, R. C. (1989). Dual attention to dynamically structured naturalistic events. *Percept. Mot. Skills*, 69:1187-1201.

Theeuwes, J. (1994). Stimulus-driven capture and attentional set: Selective search for color and visual abrupt onsets. *J. Exp. Psych. Hum. Percept. Perf.*, 20:799-806.

Treisman, A. M. (1964). Monitoring and storage of irrelevant messages in selective attention. *J. Verbal Learn. Verbal Behav.*, 3:449-459.

Treisman, A. M. (1960). Contextual cues in selective listening. *Q. J. Exp. Psych.*, 12:242-248.

Treisman, A. M., and Gelade, G. (1980). A feature-integration theory of attention. *Cog. Psych.*, 12:97-136.

Watson, D. G., and Humphreys, G. W. (1997). Visual marking: Prioritizing selection for new objects by top-down attentional inhibition of old objects. *Psych. Rev.*, 104:90-122.

Watson, D. G., and Humphreys, G. W. (1998). Visual marking of moving objects: A role for top-down feature-based inhibition in selection. *J. Exp. Psych.*, 24:46-962.

Yantis, S. (1993). Stimulus-driven attentional capture and attentional control settings. *J. Exp. Psych. Hum. Percept. Perf.*, 19:676-681.

Yantis, S. (1998). Control of visual attention. In H. Pashler (Ed.), *Attention*, pp. 223-256. East Sussex, UK: Psychology Press.

Yantis, S., and Hillstrom, A. P. (1994). Stimulus-driven attentional capture: Evidence from equiluminant visual objects. *J. Exp. Psych. Hum. Percept. Perf.*, 20:95-107.

Yantis, S., and Jonides, J. (1984). Abrupt visual onsets and selective attention: evidence from visual search. *J. Exp. Psych. Hum. Percept. Perf.*, 10:601-621.

Yantis, S., and Jonides, J. (1996). Attentional capture by abrupt onsets: New perceptual objects or visual masking? *J. Exp. Psych. Hum. Percept. Perf.*, 22:1505-1513.

11

Attention, Eye Movements, and Neurons: Linking Physiology and Behavior

Narcisse P. Bichot

Many studies have shown that attention and eye movements are guided by a common selection process. The frontal eye field plays a key role in transforming the outcome of visual selection into a command to move the eyes. Two general classes of neurons are especially important in this transformation: visually responsive neurons which provide a window into attentional selection, and movement-related neurons which mediate motor preparation. Single neurons were recorded in the frontal eye field of monkeys performing a variety of visual-search tasks that have traditionally been used in human cognitive studies of visual attention. Properties of the neural selection process in this cortical region explain behavioral characteristics of task performance. Results are discussed in terms of a visual salience map, conceivably also distributed across areas of the parietal cortex and the superior colliculus, in which object locations are tagged for behavioral relevance derived from bottom-up influences, such as conspicuousness, as well as top-down influences involving goals and knowledge.

11.1 Introduction

Vision is of primary importance to primates in gathering information about the surrounding world. However, not every part of a visual scene is processed to the same degree. Instead, we attend to objects of interest while ignoring irrelevant ones. In other words, despite our subjective feeling that we "see" everything in our field of vision, the brain's representation of the visual world is not a "faithful" one as that of a camera snapshot but one of visual information with behavioral relevance. This conclusion is supported by the intriguing literature on change blindness showing that viewers can be utterly unaware of significant changes in the information within their field of vision (Rensink, 2000; Simons, 2000).

Although psychophysical studies have greatly advanced our understanding of the phenomenology of visual selection and attention, behavioral studies ultimately provide only inferential conclusions, and the combination of behavioral and neurophysiological data is necessary for understanding the architecture and mechanisms

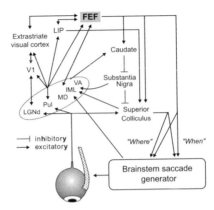

FIGURE 11.1. This diagram illustrates a simplified schematic of the connectivity between structures involved in visual selection and saccade generation. IMP – internal medullary lamina; LGNd – dorsal lateral geniculate nucleus, Pul – pulvinar, MD – mediodorsal nucleus of the thalamus, VA – ventroanterior nucleus of the thalamus, LIP – lateral intraparietal area, FEF – frontal eye field.

of human cognition (Desimone and Duncan, 1995; Schall and Bichot, 1998; Schall and Thompson, 1999). Facilitated by advances in neural recording and analysis techniques in the past two decades, there has been an explosion in the number of physiological studies aimed at investigating the neural mechanisms of visual selection. A simplified schematic of the structures involved in visual selection and saccade production is shown in Fig. 11.1.

Perhaps the most prominent aspect of the neurophysiological literature on visual selection is that neural correlates of visual selection appear to be reflected, in some form or another, in nearly all visual or visual-association areas studied to date (reviewed by Maunsell, 1995). The omnipresence of attentional modulation throughout the brain is perhaps not surprising considering that visual areas of the brain are interconnected through a complex web of both feedforward and feedback connections (Felleman and Van Essen, 1991; Salin and Bullier, 1995), possibly operating more simultaneously than once thought (Bullier and Nowak, 1995).

Neural correlates of popout or figure-ground segregation (Knierim and Van Essen, 1992; Zipser et al., 1996), and to some degree of task-dependent selection (Roelfsema et al., 1998), are found in as early as the primary visual cortex (V1). In addition, attentional modulation of neural responses have been described in extrastriate visual areas V2 and V4 (Moran and Desimone, 1985; Motter, 1994; Connor et al., 1997; Luck et al., 1997; Reynolds et al., 1999), MT and MST (Treue and Maunsell, 1999), posterior parietal cortex including area 7a and the lateral intraparietal area (LIP) (Bushnell et al., 1981; Mountcastle et al., 1981; Robinson et al., 1995; Steinmetz and Constantinidis, 1995; Shadlen and Newsome, 1996; Platt and Glimcher, 1997; Gottlieb et al., 1998), and inferior temporal cortex (Chelazzi et al., 1998). Several general principles have emerged from this extensive body of literature. Generally, responses to attended stimuli are greater than responses to unattended stimuli, accounting for the intrinsic stimulus preferences of neurons. This modulation appears to increase as one advances in the hierarchy of cortical areas starting at V1 (Moran and Desimone, 1985; Luck et al., 1997), and is typically greater if attended and anattended stimuli are in a neuron's receptive field (Luck et al., 1997; Treue and Maunsell, 1999).

This chapter focuses on neural correlates of visual selection in the frontal eye field (FEF), an area of the prefrontal cortex that plays a key role in transforming the outcome of visual processing into a command to move the eyes (reviewed by Schall and Bichot, 1998; Schall and Thompson, 1999). Evidence suggests that the FEF can be regarded as a map of the visual field in which stimulus locations are tagged for behavioral relevance derived from conspicuousness (i.e., bottom-up influences) as well as prior knowledge or expectancy (i.e., top-down influences) (Thompson and Bichot, 1999). Such a salience map has been proposed by most models of covert attention (e.g., Treisman and Gelade, 1980; Koch and Ullman, 1985; Cave and Wolfe, 1990; Olshausen et al., 1993) and overt saccade production (e.g., Findlay and Walker, 1999). Other brain structures in which a salience map may be present, such as LIP (Gottlieb et al., 1998) and the superior colliculus (reviewed by Findlay and Walker, 1999), are also discussed.

11.2 Attention and Saccades

Mechanisms of attention and visual selection have been extensively studied in humans using psychophysical measures of accuracy or reaction time. Two general categories of factors play a role in what parts of a visual scene are processed preferentially: bottom-up factors and top-down factors. Bottom-up factors are derived from the conspicuousness of features or objects in a visual scene. In other words, objects that are different from those surrounding them are preferentially selected (e.g., a bright spot in the dark or a red square among green squares). Top-down factors, on the other hand, are derived from the goals of the viewer (e.g., searching for a target letter among distractor letters). In a majority of these behavioral studies on visual attention, subjects were required to withhold saccades, and when they were not, eye movement data were usually ignored. On the other hand, the selective nature of gaze behavior has been shown elegantly by Yarbus (1967), among other researchers (see Viviani, 1990, for a review), by recording the eye movements of subjects viewing natural images. Gaze, like attention, focuses on parts of a visual scene that are conspicuous or that are informative with respect to the viewer's goals. The same type of selective visual behavior is also observed in non-human primates such as macaque monkeys (Keating and Keating, 1993; Burman and Segraves, 1994).

Although the relation between attention and eye movements is not obligatory, more recent research indicates that a common visual selection mechanism governs both covert and overt orienting. One line of behavioral evidence for the strong functional link between attention and eye movements is that perceptual performance is best when attention and saccade are directed to the same object. For example, using a paradigm designed to eliminate methodological problems in earlier studies, Shepherd et al. (1986) showed that manual reaction time in detecting a stimulus is shorter if its location coincided with the target of a saccade than when saccade target and detection stimulus were presented in opposite hemifields. Using similar dual-task paradigms, later studies by Hoffman and Subramaniam (1995) and Kowler et al. (1995) also showed that perceptual performance is best when the saccade is directed toward the stimulus to be detected or identified. These studies also provided the first evidence that the mechanisms of visual attention and saccade programming cannot be dissociated. In a subsequent study, Deubel and Schneider (1996) showed similar results and that discrimination performance is near chance level even for stimuli neighboring the saccade target. Furthermore, they showed that even when subjects knew beforehand the position of the discrimination target, it was not possible to attend to the discrimination target while programming a saccade to a neighboring target. These data further support an obligatory coupling between attention and saccades.

Other behavioral evidence suggesting a natural linkage between attention and saccades comes from studies describing similar effects on overt and covert orienting. As mentioned earlier, bottom-up factors clearly influence visual selection. Conspicuous stimuli are selected both covertly (Theeuwes, 1991) and overtly (Theeuwes et al., 1998). In addition, top-down factors are equally important in visual selection. For example, performance during search for an oddball item improves with repetition of target and distractor features across trials (Maljkovic and Nakayama, 1994). Similar short-term priming effects have been observed on eye movements (Bichot and Schall, 1999a; McPeek et al., 1999). Furthermore, cognitive strategies can prevent both covert (Bacon and Egeth, 1994) and overt (Bichot et al., 1996; Nodine et al., 1996) selection of conspicuous stimuli. Finally, in a visual search for a target defined by a combination of elementary features such as color and shape, distractor stimuli that resemble the target are more likely to be selected both covertly (Kim and Cave, 1995) and overtly (Williams, 1967; Findlay, 1997; Zelinsky and Sheinberg, 1997; Motter and Belky, 1998; Bichot and Schall, 1999a) than are distractors that do not resemble the target.

Neurophysiological studies also support the view that attention and eye movements are functionally related. Brain-imaging studies have shown that common regions in the human parietal and frontal cortex are activated in association with attention and saccade tasks (Nobre et al., 1997; Corbetta, 1998). A study by Kustov and Robinson (1996) presented both intracortical microstimulation and single-neuron recording data from the superior colliculus, a structure central to saccade production, showing that attentional shifts are associated with the preparation of eye movements. When monkeys were cued either exogenously or endogenously to attend to a location, the eye movements evoked by stimulation of the superior colliculus deviated toward the attended location. Also, "build-up" neurons in the superior colliculus were activated during both exogenously and endogenously cued attention shifts in the absence of eye movements.

Altogether, these results suggest that covert orienting may be little more than a state of visual selection without activating motor circuitry to produce overt orienting. This conclusion is consistent with the premotor theory of attention proposing that the attentional system has evolved as part of the motor systems and is part of the premotor processing of the brain (Rizzolatti et al., 1994).

11.3   Frontal Eye Field

The frontal eye field (FEF), located on the rostral bank of the arcuate sulcus, sits at the interface between vision and eye movement production. Consequently, the FEF has two aspects, one sensory and one motor. Most research on the FEF has emphasized its motor aspect, and indeed, its involvement in the generation of saccadic eye movements is universally accepted. In fact, the FEF is defined as the region of the frontal cortex from which saccades are elicited with currents of less than 50 μA (Bruce et al., 1985). The influence of the FEF on saccade production is mediated by layer 5 'movement' neurons that are active specifically before and during saccades (Bruce and Goldberg, 1985; Hanes and Schall, 1996), and that project to the superior colliculus (Segraves and Goldberg, 1987), as well as parts of the brain stem saccade-generating circuit (Segraves, 1992). Recent studies have demonstrated a profound incapacity of monkeys to generate eye movements following reversible inactivation of the FEF (Dias et al., 1995; Sommer and Tehovnik, 1997), complementing earlier observations that ablation of the FEF causes an initial severe impairment in saccade production that recovers over time in some aspects (Schiller et al., 1987), but not others (Schiller and Chou, 1998). Furthermore, although monkeys can recover from ablation of FEF or the superior colliculus alone, these two structures combined appear to account for all of the ability of primates to generate saccades as combined bilateral lesions of the FEF and superior colliculus result in the permanent loss of saccades (Schiller et al., 1980). However, Hanes and Wurtz (1999) have shown that saccades cannot be elicited by stimulating the FEF even with suprathreshold currents if topographically matched portions of the superior colliculus are inactivated. These results suggest that the contribution of these two structures to saccade production in the intact brain is not as parallel as once believed. This finding is consistent with the working hypothesis that while the FEF is more critical for voluntary, visually guided saccades, the superior colliculus is responsible for reflexive, orienting saccades, and in general, the low-level control of saccades (Schall, 1991a).

Surprisingly, the role of the FEF in visual selection has been, for the most part, ignored despite equally compelling evidence for the visual aspect of this area. Roughly, half of the neurons in the FEF have visual responses (Mohler et al., 1973; Bruce and Goldberg, 1985; Schall, 1991b). These visual responses are mediated by massive converging input from extrastriate visual areas of both the dorsal (or "where") and ventral (or "what") streams (Baizer et al., 1991; Schall et al., 1995a). In fact, a mathematical analysis of the connectivity between cortical areas indicates that the FEF is one of the highest points of convergence of dorsal and ventral stream visual information in the brain (Jouve et al., 1998). The connections between extrastriate visual areas and the FEF are reciprocal and topographically organized (Schall et al., 1995a). The ventrolateral FEF, involved in the production of short-amplitude saccades, receives visual afferents from the central field representation of retinotopically organized areas (e.g., MT and V4), from areas that emphasize central vision in the inferior temporal cortex (e.g., TEO and caudal TE), as well as areas in parietal cortex that have little retinotopic order (e.g., LIP). In contrast, the dorsomedial FEF, involved in the production of longer-amplitude saccades, receives visual afferents from the peripheral field representation of retinotopically organized areas, and from areas that emphasize peripheral vision (e.g., PO and MSTd). In addition, the FEF receives input from areas in the prefrontal cortex (Stanton et al., 1993).

11.4   Bottom-Up Influences on Visual Selection

Bottom-up visual selection refers to the allocation of attention based solely on image properties. For example, a stimulus that differs in one or more visual attributes (e.g., brightness, color, orientation, shape, or direction of motion) from neighboring stimuli is said to "pop out" and is likely to be attended and fixated. Such conspicuous stimuli attract attention automatically, and in some instances, despite the fact that they are irrelevant to the viewer's goals (Theeuwes, 1991). Visual search for a pop-out target is said to be automatic or "effortless" as evidenced by little or no effect of increasing the number of distracting items on the time it takes to detect the target (e.g., Treisman and Gelade, 1980). A study by Bichot and Schall (1999a) shows that monkeys perform pop-out visual searches similar to humans in that the latency or accuracy of their saccades was not affected by the number of distracting elements (Fig. 11.2). Furthermore, errant saccades were directed more often to distractors near the oddball target than to a distractor at any other location, consistent with the strong capture of attention by conspicuous stimuli.

Schall and colleagues (Schall and Hanes, 1993; Schall et al., 1995b) have investigated how the brain selects conspicuous stimuli in monkeys trained to make a saccade to a target defined by a unique feature. Recordings in monkeys trained to fixate the oddball target in complementary search arrays (e.g., red item among green items and vice versa) (Fig. 11.3) have shown that the initial activity of most visually responsive neurons in the FEF does not discriminate whether the target or only distractors of the search array appeared in their receptive field (Fig. 11.3). This observation is not surprising because earlier work has shown that visual responses in the FEF are not selective for stimulus properties (Mohler et al., 1973). However, before saccades were generated, the activity of FEF neurons evolved to signal the location of the oddball target regardless of the visual feature that distinguished it from distractors. Preliminary results suggest that a similar selection of an oddball stimulus takes place in area 7a of the posterior parietal cortex (Constantinidis and

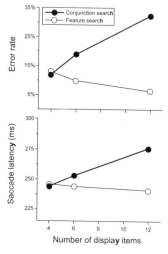

FIGURE 11.2. Error rates and saccade latency as a function of the number of items in the display during conjunction search (filled circles, thick lines) and feature search (outline circles, thin lines) performed by macaque monkeys. (Modified from Bichot and Schall, 1999a.)

Steinmetz, 1996).

The target-selection signal observed in the FEF has a number of interesting properties (Schall et al., 1995b). First, the initial response to the search array was consistently attenuated relative to the initial response to the target presented alone (e.g., detection trials). This suppression may reflect the operation of antagonistic suppressive connections, or it may arise from the uncertainty of the target position in the search display as suggested by a recent study in the superior colliculus (Basso and Wurtz, 1998). Second, target selection appears to be mediated by the suppression of the activity evoked by distractors, consistent with the findings of a recent study measuring attentional allocation at target and distractor locations during a pop-out search (Cepeda et al., 1998). However, other studies have shown that both distractor suppression and target facilitation play a role in visual selection (e.g., Maljkovic and Nakayama, 1994), and further experiments are necessary to tease apart the relative contribution of these two mechanisms to the selection process observed in the FEF. Third, for some neurons this suppression was stronger when the target flanked the receptive field. This flanking suppression may be important in reducing the probability of saccades to distractors in the receptive field, and may underlie the flanking suppression observed in psychophysical measurements of attention by Cave and Zimmerman (1997). A biologically-plausible model of attention by Tsotsos et al. (1995) shows how such flanking suppression can arise in the visual system.

A central idea in experimental psychology is that reaction times are composed of stages of processing (reviewed by Meyer et al., 1988). Using an analysis adapted from signal-detection theory, Thompson et al. (1996) examined whether the out-

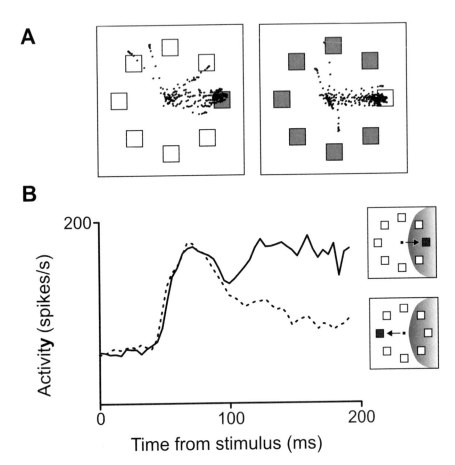

FIGURE 11.3. (a) Patterns of gaze shift of one monkey trained to make saccades to oddball targets regardless of target and distractor properties. Black squares represent actual red squares, and white squares represent actual green squares. (b) Neural activity of one FEF visually responsive neuron from that monkey during pop-out search. The average discharge of the neuron when the oddball fell in its receptive field (gray shaded area in search display) is indicated by the solid spike-density function, and the average discharge of the neuron when only distractors fell in its receptive field is indicated by the dotted spike-density function. In this and other figures of neural responses, activity is aligned on stimulus presentation at time zero, only spikes that occurred before saccade initiation are used in the analyses, and spike-density functions are plotted up to the mean saccade latency. The arrow in the search displays indicates the saccade to the target. (Modified from Bichot et al., 1996 and Thompson et al., 1996.)

come of visual processing indexed by visually responsive FEF neurons accounts for the well-known variability of saccadic reaction times (reviewed by Carpenter, 1988). For a majority of these neurons, discrimination occurred at a fairly constant interval after stimulus presentation (Fig. 11.4) and did not predict when a saccade would be initiated. Thus, the additional delay and variability observed in reaction times is introduced by a postperceptual stage of response preparation, as described by Hanes and Schall (1996).

To determine whether the neural selection process expressed in the FEF is contingent on saccade planning, Thompson et al. (1997) recorded single-unit activity during no-go visual-search trials. In go trials, reward was contingent on executing a saccade to the oddball stimulus of the search array, whereas in no-go trials reward was contingent on withholding saccades. Saccade planning was effectively discouraged, as indicated by the eye movements made by monkeys after correct no-go trials as well as an attenuation of visual responses in no-go trials as compared with go trials (see Goldberg and Bushnell, 1981). During no-go trials, the activity of the majority of neurons evolved to signal the location of the oddball stimulus (Fig. 11.5). Furthermore, the degree and time course of the stimulus-discrimination process observed in no-go trials were not different from that observed in go trials. These results suggest that the FEF may play a role in covert attentional selection, a conclusion supported by recent brain imaging studies showing that the FEF is activated during both attention and saccade tasks (Nobre et al., 1997; Corbetta et al., 1998).

Although singletons usually attract attention automatically, stimuli presented as sudden onsets appear to capture attention inevitably (reviewed by Yantis, 1996). A study by Gottlieb et al. (1998) provides a neural account of attentional capture by flashed stimuli. They studied visual search in the lateral intraparietal area (LIP) with a very interesting twist: they used a "stable-stimulus" paradigm in which stimuli that remained on the screen were brought into the receptive field of LIP neurons by saccades. Under these circumstances, LIP neurons that responded strongly to the sudden appearance of stimuli in their receptive field exhibited weak or no responses to stimuli brought into their receptive field by saccades unless the stimuli were behaviorally relevant. Further experiments are needed to determine whether the strong early activation in the FEF in response to the presentation of a search array is also related to the automatic attentional capture of sudden onsets. A study by Burman and Segraves (1994) in which visual responses were recorded in the FEF while monkeys freely scanned natural images supports this possibility.

11.5   Top-Down Influences on Visual Selection

Top-down visual selection refers to the allocation of attention derived from internal influences such as knowledge, expectations, and goals of the viewer. Although attention is commonly attracted to conspicuous stimuli, knowledge of what to look for is equally important in determining what parts of a visual image will receive

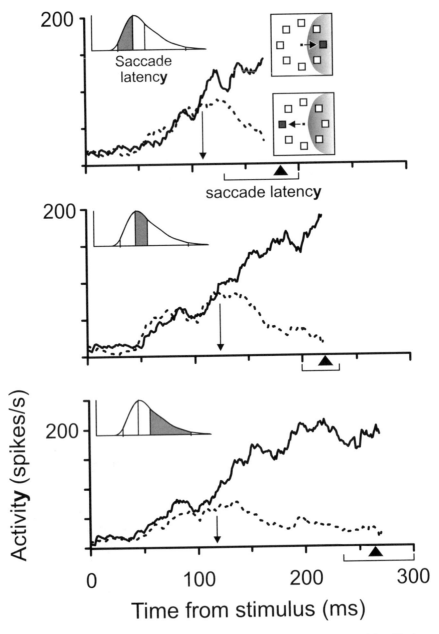

FIGURE 11.4. Time course of neural selection for one FEF neuron during trials with short (top panel), medium (middle panel), and long (bottom panel) saccade latencies. A schematic of the distribution of saccade latencies as well as the latency range used is shown in the upper left-hand corner of each panel. The range of saccade latencies are also shown under the abscissa, with the arrowhead marking the average latency for the range. Clearly, the time at which the responses to the target (solid lines) and to the distractors (dotted lines) become different (arrows) does not predict the saccadic latencies. (Modified from Thompson et al., 1996.)

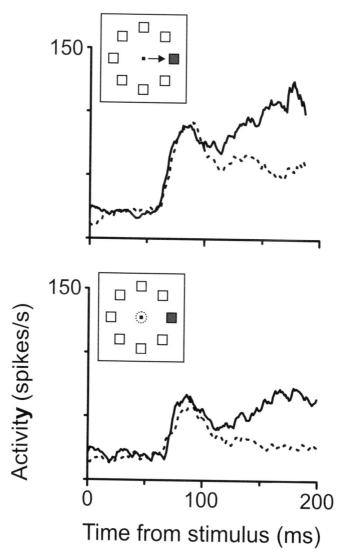

FIGURE 11.5. Activity of one FEF neuron when the monkey was instructed to make a saccade to the oddball (go search, upper panel) and when the monkey was instructed to withhold saccades to the oddball and maintain fixation at the central spot (no-go search, lower panel). This neuron discriminates the target (solid lines) from the distractors (dotted lines) whether or not a saccade to the oddball is generated. (Modified from Thompson et al., 1997.)

preferential treatment. Behavioral studies have shown that search is facilitated by the repetition of target and distractor properties (Maljkovic and Nakayama, 1994; Bichot and Schall, 1999a), target location (Maljkovic and Nakayama, 1996), and display layout (Chun and Jiang, 1998). In fact, cognitive strategies can, in some instances, override the effects of conspicuousness (Bacon and Egeth, 1994). For example, experts are more likely than novices to ignore conspicuous but non-informative elements of a visual image from their area of expertise (Nodine et al., 1996; see also Chapman and Underwood, 1998).

Bichot et al. (1996) found that the neural-selection process in the FEF can be modified profoundly by cognitive strategies. Monkeys trained to search for an odd-ball regardless of the particular feature that defines it (e.g., Fig. 11.3) generalized a strategy of shifting their gaze according to visual conspicuousness. As mentioned earlier, FEF neurons of such monkeys do not exhibit feature selectivity, but their activity evolves to signal the location of the oddball stimulus (Fig. 11.3) (e.g., Schall et al., 1995b; Thompson et al., 1996). In contrast, monkeys given exclusive experience with one visual-search array (e.g., red among green) adopted a strategy of ignoring stimuli with the distractor feature, even if those same stimuli became the oddball target in the complementary visual-search array (e.g., green among red) presented occasionally (Fig. 11.6). In monkeys using this strategy, about half of FEF neurons exhibited a suppressed response to the learned distractor as soon as they responded (Fig. 11.6). In other words, FEF neurons exhibited an apparent feature selectivity in their initial response unlike what had been observed before in this area. This study shows how cognitive strategies can dramatically affect gaze behavior and the underlying neural-selection signals even when the target can be easily detected.

In the study by Bichot et al. (1996), the search process was affected by ex-pectation of stimulus properties. More recent experiments by Basso and Wurtz (1998) have examined the effects of target-location uncertainty on the activity of superior colliculus neurons. In one experiment, the number of possible targets was changed. In this task, monkeys were presented with 1, 2, 4, or 8 identical stimuli, and were required to make a saccade to the stimulus that dimmed after a certain delay. The initial activity of buildup neurons in this structure was modulated by target probability as neurons responded the strongest when only one stimulus was presented and target location was most predictable, and their response gradually decreased as potential target locations increased. This modulation persisted until the dimming of the target stimulus, after which time the responses were the same in all stimulus configurations. Fixation and burst neurons whose responses are more closely linked to the impending saccade were largely unaffected by changes in target probability. However, the modulation just described can also be explained by the operation of antagonistic suppressive connections whose effects would be more pronounced as stimulus density is increased. To rule out this alternative, Basso and Wurtz (1998) kept the visual display constant and varied over trials the probability that a particular stimulus would be the target. Consistent with the hy-pothesis that target uncertainty modulates superior colliculus neurons, responses were greater when the target location remained constant across trials.

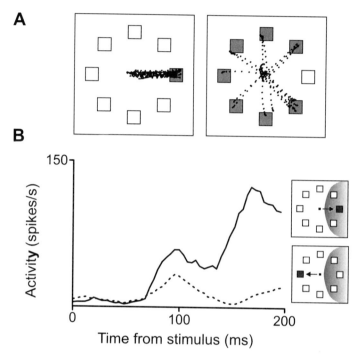

FIGURE 11.6. (a) Patterns of gaze shift of one monkey trained to make saccades with only one popout search display (left-hand panel). When occasionally presented with the complementary array, the monkey shifted gaze to the distractors with the learned target attribute, instead of the oddball stimulus with the learned distractor attribute (right-hand panel). (b) Neural activity of one FEF visually responsive neuron from that monkey during search with the learned display. This neuron responded differentially to the target (solid line) and distractors (dotted line) in its receptive field as soon as it became active (compare to non-selective initial response of neuron in Fig. 11.3). (Modified from Bichot et al., 1996.)

In many situations, objects of interest cannot be located based on intrinsic s-timulus properties alone. When objects are defined by combinations of features, knowledge and vision must collaborate to guide gaze because memory is required to locate an object with particular features. To explore neural mechanisms of visual selection when knowledge is required to find the target, Bichot and Schall (1999a,b) trained monkeys to perform a conjunction visual search in which the target during each recording session was specified as one combination of color and shape, and distractors were formed by other possible combinations.

The dichotomy between popout search and conjunction search reported in an early experiment by Treisman and Gelade (1980) played a critical role in the development of theories of visual search and attention. While popout search was effortless and could seemingly be performed 'preattentively', detecting the target during conjunction search was attentionally demanding as evidenced by significant increases in reaction time with increasing number of distractors. Subsequent experiments, however, showed that conjunction search can be performed efficiently, as reflected by smaller effects of set size on target detection (Nakayama and Silverman, 1986; Wolfe et al., 1989). These findings led to the development of models of visual search in which selection is guided by the similarity between the target and distractors (Duncan and Humphreys, 1989), most likely through parallel processing of the individual features that define the conjunction stimuli (Cave and Wolfe, 1990; Treisman and Sato, 1990).

Bichot and Schall (1999a) showed that monkeys can perform conjunction search as efficiently as humans, although a conjunction search was still clearly more demanding than a pop-out search (Fig. 11.2). Furthermore, Bichot and Schall (1999b) observed two influences on gaze behavior and the neural selection process expressed in the FEF. First, in contrast to popout search, errant saccades during conjunction search were guided more by similarity than proximity to the target, landing on distractors that shared a target feature significantly more often than on a distractor that shared none (Fig. 11.7, right). Similar observations have been made with human observers during both saccade (Findlay, 1997) and attention (Kim and Cave, 1995) tasks. The neural basis of this gaze behavior was evident in the FEF: even when monkeys successfully shifted gaze to the target, the neural representation of distractors similar to the target was stronger than that of dissimilar distractors, with the maximal activation elicited by the target in the receptive field (Fig. 11.7, left). This finding provides the first neurophysiological evidence that efficient selection during conjunction search is accomplished based on visual similarity, most likely through parallel processing of objects based on their elementary features. Furthermore, these data lend additional support to the hypothesis that the FEF participates in visual and not just motor selection because the selection process was influenced by visual similarity to the target rather than just the response being produced – that is, a saccade to the target.

The second influence observed on gaze behavior and the neural selection was rather unexpected. The history of target properties across sessions affected visual selection: in single sessions, there was an increased tendency of saccades to a distractor that had been the target during the previous session. This effect was

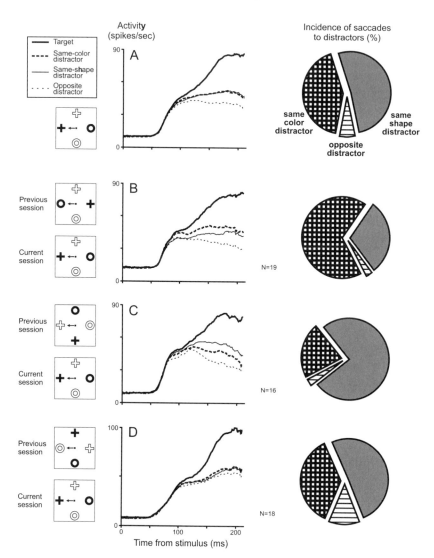

FIGURE 11.7. Neural selection in correct trials (left) and incidence of saccades to distractors (right) during conjunction visual search. Panels on the left show the pooled response of a population of FEF neurons across all conditions (a), and as a function of the target properties in the previous session: same color as current target (b), same shape as current target (c), and opposite of current target (d). The number (N) of neurons that contributed to each of the lower three plots is indicated. The top panel represents an average of the spike densities obtained in the three conditions. Responses to the target are indicated with thick solid lines, to the distractor that shared the target color with thick dotted lines, to the distractor that shared the target shape with thin solid lines, and to the distractor that shared no target features with thin dotted lines. Average responses were calculated during correct trials with both four and six stimulus displays.

observed when the target during the previous session was the distractor that shared the target color in the current session (Fig. 11.7), the distractor that shared the target shape in the current session (Fig. 11.7), and the distractor that shared no target feature in the current session (Fig. 11.7). Again, the neural basis of this effect was evident in the FEF with a relative increase in the neural representation of the distractor that was the target during the previous session (Figs. 11.7). This influence may be related to the well-known perceptual priming observed during pop-out search with human observers. However, in monkeys (Bichot and Schall, 1999a) as in humans (Maljkovic and Nakayama, 1994), this effect lasts for about ten trials, or less than a minute. In contrast, the effect observed during conjunction search expressed itself across sessions at least a day apart and persisted throughout a session. Hence, this novel influence of experience on performance may be a more enduring manifestation of the short-term priming described in previous studies.

A study by Gottlieb et al. (1998) in LIP also studied top-down visual selection, although their experimental paradigm differed in that search stimuli with various colors and shapes were present on the screen at all times. The responses of LIP neurons were studied by bringing one element of the search array into the receptive field by making monkeys first saccade to a fixation point at the center of the search array. Monkeys then made a saccade to the target stimulus instructed during either the first or the second fixation duration. As mentioned earlier, under these circumstances, LIP neurons exhibit weak or no responses to the stimulus brought into their receptive field by the first saccade. However, before the saccade was made to the target, neurons were activated if the stimulus in their receptive field was the target. Thus, it appears that stimuli are weakly or not represented in LIP unless they are behaviorally relevant, whether because they draw attention automatically, as in the case of sudden-onset stimuli, or because they are made relevant by instructions.

Shadlen and Newsome (1996) have extended demonstrations of the neural correlates of visual selection in the parietal cortex to the motion feature dimension. They recorded from neurons in area LIP while monkeys performed a match-to-sample task in which they were required to make a saccade (after a given delay) to one of two targets that matched the direction of coherent motion that appeared within a circular aperture above fixation. They investigated how well the activity of LIP neurons predicted the animals' choice (as determined by saccadic responses) and found that the predictive power of LIP neurons improved with time and stimulus strength. For stronger coherences, predictive activity developed more quickly and reached a higher level by the end of the period during which the sample stimulus was presented. The dependence of the signals carried by LIP neurons on stimulus properties argues, as in the case of FEF, against such signals being purely motor-related, as the accuracy of saccades is not determined by the coherence levels of the sample stimulus. Kim and Shadlen (1999) have made similar observations in the dorsolateral prefrontal cortex, including the FEF, and Horwitz and Newsome (1999) have shown that while some superior colliculus neurons are involved in movement preparation and execution, others carry feature-selective visual signals, consistent with a role in target selection.

11.6   Conclusions

This chapter described neural activity in structures such as the FEF, LIP, and superior colliculus that can be viewed as maps of the visual field in which the behavioral relevance of stimuli is represented, whether relevance is derived from the intrinsic properties of the stimuli or from the viewer's knowledge and goals. As mentioned, neurons in these structures do not typically exhibit selectivity for stimulus features such as color, shape, or direction of motion, and thus are ideally suited to respond preferentially to stimuli according to their behavioral relevance regardless of their features. An obvious question is then where visual properties are analyzed and stimuli selected. The properties that distinguish stimuli are represented in appropriate areas of the visual cortex in which concomitant selection occurs (e.g., Luck et al., 1997; Chelazzi et al., 1998; Treue and Maunsell, 1999) and that project to a master visual salience map that likely exists in the FEF, among other candidate structures. As reviewed, the connectivity of the visual system supports this hypothesis.

Consider for example the operation of such a salience map in FEF during a color/shape conjunction search. Properties of the stimuli are processed in extrastriate visual areas by populations of neurons that are selective for different stimulus attributes such as color and shape. In the color map, objects with the target color are selected. Similarly, in the shape map objects with the target shape are selected. Note, however, that other visual processing schemes are possible in which, for example, selection is accomplished by neurons that are selective for combinations of features. Evidence has shown that some neurons in visual areas TEO and caudal TE are selective for combinations of elementary features such as color and shape (Komatsu and Ideura, 1993). Finally, a top-down memory signal of target properties is necessary to locate the target. Such top-down influence is likely exerted by areas of the prefrontal cortex that have been implicated in working memory (Mishkin and Manning, 1978; Miller et al., 1996).

As a result of converging projections onto the salience map, the location of the target receives the most activation, the locations of distractors similar to the target receive intermediate activation, and the location of the distractor that shares no features with the target receives the least activation. Then, possibly as a result of winner-take-all competition, neurons involved in saccade preparation and execution can be activated, leading to a gaze shift. Of course, if activations in the salience map were always this perfect, observers would correctly locate the target on every trial. Models of visual search have solved this issue by assuming that activations are noisy whether this noise is intrinsic to the visual system (Cave and Wolfe, 1990) or arises from bottom-up selection mechanisms still being active (Cave et al., in press).

Although it is likely that the selection observed in the FEF reflects signals generated in the extrastriate visual cortex, we should not rule out the possibility that signals in the FEF can influence neural processing in the visual cortex via feedback projections (e.g., Schall et al., 1995a). Regardless, results suggest that the FEF can be viewed as a window into visual selection that takes place across many structures of the brain, and thus provides an ideal area in which to test theories on the mechanisms of visual selection as well as response production.

Acknowledgments

I would like to thank Drs. Robert Desimone, Jeffrey Schall, and Kirk Thompson for their helpful comments. This work was supported by National Eye Institute grants RO1 EY08890 to Dr. Jeffrey Schall and P30 EY08126 and T32 EY07135 to the Vanderbilt Vision Research Center, and by the National Institute of Mental Health Intramural Research Program.

References

Bacon, W. F. and Egeth, H. E. (1994). Overriding stimulus-driven attentional capture. *Percept. and Psychophys.*, 55:485-496.

Baizer, J. S., Ungerleider, L. G. and Desimone, R. (1991). Organization of visual inputs to the inferior temporal and posterior parietal cortex in macaques. *J. Neurosci.*, 11:168-190.

Basso, M. A. and Wurtz, R. H. (1998). Modulation of neuronal activity in superior colliculus by changes in target probability. *J. Neurosci.*, 18:7519-7534

Bichot, N. P. and Schall, J. D. (1999a). Saccade target selection in macaque during feature and conjunction visual search. *Vis. Neurosci.*, 16:81-89.

Bichot, N. P. and Schall, J. D. (1999b). Effects of similarity and history on neural mechanisms of visual selection. *Nature Neurosci.* 2:549-554.

Bichot, N. P., Schall, J. D., and Thompson, K. G. (1996). Visual feature selectivity in frontal eye fields induced by experience in mature macaques. *Nature*, 381:697-699.

Bruce, C. J., and Goldberg, M. E. (1985). Primate frontal eye fields I: Single neurons discharging before saccades. *J. Neurophysiol.* 53: 603-635.

Bruce, C. J., Goldberg, M. E., Bushnell. C., and Stanton, G. B. (1985). Primate frontal eye fields II: Physiological and anatomical correlates of electrically evoked eye movements. *J. Neurophysiol.* 54:714-734.

Bullier, J. and Nowak, L. G. (1995). Parallel versus serial processing: New vistas on the distributed organization of the visual system. *Cur. Opin. Neurobiol.*, 5:497-503.

Burman, D. D. and Segraves, M. A. (1994). Primate frontal eye field activity during natural scanning eye movements. *J. Neurophysiol.*, 71:1266-1271.

Bushnell, M. C., Goldberg, M. E. and Robinson, D. L. (1981). Behavioral enhancement of visual responses in monkey cerebral cortex. I. Modulation in posterior parietal cortex related to selective visual attention. *J. Neurophysiol.* 46: 755-772.

Carpenter, R. H. S. (1988). *Movements of the Eyes.* London: Pion Press.

Cave, K. R., Kim, M.-S., Bichot, N. P. and Sobel, K. V. (in press). Visual selection within a hierarchical network: the FeatureGate model. *Cog. Psych.*.

Cave, K. R. and Wolfe, J. M. (1990). Modeling the role of parallel processing in visual search. *Cog. Psych.*, 22:225-271.

Cave, K. R. and Zimmerman, J. M. (1997). Flexibility in spatial attention before and after practice. *Psychol. Sci.*, 8:399-403.

Cepeda, N. J., Cave, K. R., Bichot, N. P., and Kim, M.-S. (1998). Spatial s-election via feature-driven inhibition of distractor locations. *Percept. and Psychophys.*, 60:727-746.

Chapman, P. and Underwood, G. (1998). Visual search of driving situations: danger and experience. *Percept.*, 27:951-964.

Chelazzi, L., Duncan, J., Miller, E. K., and Desimone, R. (1998). Responses of neurons in inferior temporal cortex during memory-guided visual search. *J. Neurophysiol.*, 80:2918-2940.

Chun, M. M., and Jiang, Y. (1998). Contextual cueing: Implicit learning and memory of visual context guides spatial attention. *Cog. Psych.*, 36:28-71.

Connor, C. E., Preddie, D. C., Gallant, J. L., and Van Essen, D. C. (1997). Spatial attention effects in macaque area V4. *J. Neurosci.* 17:3201-3214.

Constantinidis, C. and Steinmetz, M. A. (1996). Neuronal responses in area 7a to stimuli that attract attention. *Soc. Neurosci. Abst.*, 22:1198.

Corbetta, M. (1998). Frontoparietal cortical networks for directing attention and the eye to visual location: identical, independent, or overlapping neural systems? *Proc. Nat. Acad. Sci. USA*, 95:831-838.

Desimone, R. and Duncan, J. (1995). Neural mechanisms of selective visual attention. *Ann. Rev. Neurosci.*, 18:193-222.

Deubel, H. and Schneider, W. X. (1996). Saccade target selection and object recognition: Evidence for a common attentional mechanism. *Vis. Res.*, 36:1827-1837.

Dias, E. C., Kiesau, M., and Segraves, M. A. (1995). Acute activation and inactivation of macaque frontal eye field with GABA-related drugs. *J. Neurophysiol.*, 74:2744-2748.

Duncan, J. and Humphreys, G. W. (1989). Visual search and stimulus similarity. *Psych. Rev.*, 96:433-458.

Felleman, D. J. and Van Essen, D. C. (1991). Distributed hierarchical processing in the primate cerebral cortex. *Cerebral Cortex*, 1:1-47.

Findlay, J. M. (1997). Saccade target selection during visual search. *Vis. Res.*, 37:617-631.

Findlay, J. M. and Walker, R. (1999). A model of saccade generation based on parallel processing and competitive inhibition. *Behav. Brain Sci.*, 22:661-721.

Goldberg, M. E. and Bushnell, M. C. (1981). Behavioral enhancement of visual responses in monkey cerebral cortex. II. Modulation in frontal eye fields specifically related to saccades. *J. Neurophysiol.*, 46:773-787.

Gottlieb, J. P., Kusunoki, M., and Goldberg, M. E. (1998). The representation of visual salience in monkey parietal cortex. *Nature*, 391:481-484.

Hanes, D. P. and Schall, J. D. (1996). Neural control of voluntary movement initiation. *Science*, 274:427-430.

Hanes, D. P. and Wurtz, R. H. (1999). Effects of superior colliculus inactivation on saccades evoked by frontal eye field stimulation. *Soc. Neurosci. Abstr.*, 25:805.

Hoffman, J. E. and Subramaniam, B. (1995). The role of visual attention in saccadic eye movements. *Percept.and Psychophys.*, 57:787-795.

Horwitz, G. D. and Newsome, W. T. (1999). Separate signals for target selection and movement specification in the superior colliculus. *Science*, 284:1158-1161.

Jouve, B., Rosenstiehl, P. and Imbert, M. (1998). A mathematical approach to the connectivity between the cortical visual areas of the macaque monkey. *Cerebral Cortex*, 8:28-39.

Keating, C. F. and Keating, E. G. (1993). Monkeys and mug shots: Cues used by rhesus monkeys (*Macacca Mulatta*) to recognize a human face. *J. Compar. Psych.*, 107:131- 139.

Kim, M.-S. and Cave, K. R. (1995). Spatial attention in visual search for features and feature conjunctions. *Psych. Sci.*, 6:376-380.

Kim, J. N. and Shadlen, M. N. (1999). Neural correlates of a decision in the dorsolateral prefrontal cortex of the macaque. *Nature Neurosci.*, 2:176-185.

Knierim, J. J. and Van Essen, D. C. (1992). Neuronal responses to static texture patterns in area V1 of the alert macaque monkey. *J. Neurophysiol.*, 67:961-980.

Koch, C. and Ullman, S. (1985). Shifts in selective visual attention: towards the underlying neural circuitry. *Hum. Neurobiol.*, 4:219-227.

Komatsu, H. and Ideura, Y. (1993). Relationships between color, shape, and pattern selectivities of neurons in the inferior temporal cortex of the monkey. *J. Neurophysiol.*, 70:677-694.

Kowler, E., Anderson, E., Dosher, B.,and Blaser, E. (1995). The role of attention in the programming of saccades. *Vis. Res.*, 35:1897-1916.

Kustov, A. A. and Robinson, D. L. (1996). Shared neural control of attentional shifts and eye movements. *Nature*, 384:74-77.

Luck, S. J., Chelazzi, L., Hillyard, S. A. and Desimone, R. (1997). Neural mechanisms of spatial selective attention in areas V1, V2, and V4 of macaque visual cortex. *J. Neurophysiol.*, 77:24-42.

Maljkovic, V. and Nakayama, K. (1994). Priming of pop-out: I. Role of features. *Mem. Cog.*, 22:657-672.

Maljkovic, V. and Nakayama, K. (1996). Priming of pop-out: II. Role of position. *Percept. Psychophys.*, 58:977-991.

Maunsell, J. H. R. (1995). The brain's visual world: representation of visual targets in cerebral cortex. *Science*, 270:764-769.

McPeek, R. M., Maljkovic, V. and Nakayama, D. (1999). Saccades require focal attention and are facilitated by a short-term memory system. *Vis. Res.*, 39:1555-1566.

Meyer, D. E., Osman, A. M., Irwin, D. E., and Yantis, S. (1988). Modern mental chronometry. *Biol. Psych.*, 26:3-67.

Miller, E. K., Erickson, C. A., and Desimone, R. (1996). Neural mechanisms of visual working memory in prefrontal cortex of the macaque. *J. Neurosci.*, 16:5154-5167.

Mishkin, M. and Manning, F. J. (1978). Non-spatial memory after selective prefrontal lesions in monkeys. *Brain Res.*, 143:313-323.

Mohler, C. W., Goldberg, M. E., and Wurtz, R. H. (1973). Visual receptive fields of frontal eye field neurons. *Brain Res.*, 61:385-389.

Moran, J. and Desimone, R. (1985). Selective attention gates visual processing in the extrastriate cortex. *Science,*, 229:782-784.

Motter, B. C. (1994). Neural correlates of attentive selection for color or luminance in extrastriate area V4. *J. Neurosci.*, 14:2178-2189.

Motter, B. C. and Belky, E. J. (1998). The guidance of eye movements during active visual search. *Vis. Res.*, 38:1805-1815.

Mountcastle, V. B., Andersen, R. A., and Motter, B. C. (1981). The influence of attentive fixation upon the excitability of light-sensitive neurons of the posterior parietal cortex. *J. Neurosci.*, 1:1218-1235.

Nakayama, K. and Silverman, G. H. (1986). Serial and parallel processing of visual feature conjunctions. *Nature*, 320:264-265.

Nobre, A. C., Sebestyen, G. N., Gitelman, D. R., Mesulam, M. M., Frackowiak, R. S. J., and Frith, C. D. (1997). Functional localization of the system for visuospatial attention using positron emission tomography. *Brain*, 120:515-533.

Nodine, C. F., Kundel, H. L., Lauver, S. C., and Toto, L. C. (1996). Nature of expertise in searching mammograms for breast masses. *Acad. Radiol.*, 3:1000-1006.

Olshausen, B. A., Anderson, C. H., and Van Essen, D. C. (1993). A neurobiological model of visual attention and invariant pattern recognition based on dynamic routing of information. *J. Neurosci.*, 13:4700-4719.

Platt, M. L. and Glimcher, P. W. (1997). Responses of intraparietal neurons to saccadic targets and visual distractors. *J. Neurophysiol.*, 78:1574-1589.

Rensink, R. (2000). Change blindness: implications for the role of attention in scene perception. In M. Jenkin and L. Harris (Eds.) *Vision and Attention* New York, NY: Springer-Verlag.

Reynolds, J. H., Chelazzi, L., and Desimone, R. (1999) Competitive mechanisms subserve attention in macaque areas V2 and V4. *J. Neurosci.*, 19:1736-1753.

Rizzolatti, G., Riggio, L., and Sheliga, B. M. (1994). Space and selective attention. In C. Umilta and M. Moscovitch (Eds.) *Attention and Performance XV*, pp. 231-265. Cambridge, MA: MIT Press.

Robinson, D. L., Bowman, E. M., and Kertzman, C. (1995). Covert orienting of attention in macaques. II. Contributions of parietal cortex. *J. Neurophysiol.*, 74:698-712.

Roelfsema P. R., Lamme, V. A. and Spekreijse, H. (1998). Object-based attention in the primary visual cortex of the macaque monkey. *Nature*, 395:376-381.

Salin, P.-A. and Bullier, J. (1995). Corticortical connections in the visual system: structure and function., *Physiol. Rev.*, 75:107-154.

Schall, J. D. (1991a). Neuronal basis of saccadic eye movements. In A. G. Leventhal (Ed.), *The Neural Basis of Visual Function*, pp. 388-442. London: Macmillan Press.

Schall, J. D. (1991b). Neuronal activity related to visually guided saccades in the frontal eye fields of rhesus monkeys: Comparison with supplementary eye fields. *J. Neurophysiol.*, 66:559-579.

Schall, J. D. and Bichot, N. P. (1998). Neural correlates of visual and motor decision processes. *Current Opinion in Neurobiol.*, 8:211-217.

Schall, J. D. and Hanes, D. P. (1993). Neural basis of saccade target selection in frontal eye field during visual search. *Nature*, 366:467-469.

Schall, J. D., Hanes, D. P., Thompson, K. G., and King, D. J. (1995a). Saccade target selection in frontal eye field of macaque. I. Visual and premovement activation. *J. Neurosci.*, 15:6905-6918.

Schall, J. D., Morel, A., King, D. J., and Bullier, J. (1995b). Topography of visual cortical afferents to frontal eye field in macaque: Functional convergence and segregation of processing streams. *J. Neurosci.*, 15:4464-4487.

Schall, J. D. and Thompson, K. G. (1999). Neural selection and control of visually guided eye movements. *Ann. Rev. Neurosci.*, 22:241-259.

Schiller, P. H. and Chou, I.-H. (1998). The effects of frontal eye field and dorsomedial frontal cortex lesions on visually guided eye movements. *Nature Neurosci.*, 1:248-253.

Schiller, P. H., Sandell, J. H., and Maunsell, J. H. R. (1987). The effect of frontal eye field and superior colliculus lesions on saccadic latencies in the rhesus monkey. *J. Neurophysiol.*, 57:1033-1049.

Schiller, P. H., True, S. D., and Conway, J. D. (1980). Deficits in eye movements following frontal eye field and superior colliculus ablations. *J. Neurophysiol.*, 44:1175-1189.

Segraves, M. A. (1992). Activity of monkey frontal eye field neurons projecting to oculomotor regions of the pons. *J. Neurophysiol.*, 68:1967-1985.

Segraves, M. A. and Goldberg, M. E. (1987). Functional properties of corticotectal neurons in the monkey's frontal eye field. *J. Neurophysiol.*, 58:1387-1419.

Shadlen, M. N. and Newsome, W. T. (1996). Motion perception: Seeing and deciding. *Proc. Nat. Acad. Sci. USA*, 93:628-633.

Shepherd, M., Findlay, J. M., and Hockey, G. R. J. (1986). The relationship between eye movements and spatial attention. *Q. J. Exp. Psych.*, 38A:475-491.

Simons, D. J. (2000). Change blindness and inattentional blindness: the role of attention and inattention in perception. In M. Jenkin and L. Harris (Eds.) *Vision and Attention*, New York, NY: Springer-Verlag.

Sommer, M. A., Tehovnik, E. J. (1997). Reversible inactivation of macaque frontal eye field. *Exp. Brain Res.*, 116:229-249.

Stanton, G. B., Bruce, C. J., and Goldberg, M. E. (1993). Topography of projections to the frontal lobe from the macaque frontal eye fields. *J. Compar. Neurol.*, 330:286-301.

Steinmetz, M. A. and Constantinidis, C. (1995). Neurophysiological evidence for a role of posterior parietal cortex in redirecting visual attention. *Cerebral Cortex*, 5:448-456.

Theeuwes, J. (1991). Cross-dimensional perceptual selectivity. *Percept. Psychophys.*, 50:184-193.

Theeuwes, J., Kramer, A. F., Hahn, S., and Irwin, D. E. (1998). Our eyes do not always go where we want them to go: capture of the eyes by new objects. *Psychol. Sci.*, 9:379-385.

Thompson, K. G., Bichot, N. P., and Schall, J. D. (1997). Dissociation of visual discrimination from saccade programming in macaque frontal eye field. *J. Neurophysiol.*, 77:1046-1050.

Thompson, K. G. and Bichot, N. P. (1999). Frontal eye field: a cortical salience map. *Behav. and Brain Sci.*, 22:699-700.

Thompson, K. G., Hanes, D. P., Bichot, N. P., and Schall, J. D. (1996). Perceptual and motor processing stages identified in the activity of macaque frontal eye field neurons during visual search. *J. Neurophysiol.*, 76:4040-4054.

Treisman, A. and Sato, S. (1990). Conjunction search revisited. *J. Exp. Psych. Hum. Percept. Perf.* 16:456-478.

Treisman, A. M. and Gelade, G. (1980). A feature-integration theory of attention. *Cog. Psych.*, 12:97-136.

Treue, S. and Maunsell, J. H. R. (1999). Effects of attention on the processing of motion in macaque middle temporal and medial superior temporal visual cortical areas. *J. Neurosci.*, 19:7591-7602.

Tsotsos, J. K., Culhane, S., Wai, W., Lai, Y., Davis, N., and Nuflo, F. (1995). Modeling visual attention via selective tuning. *Artif. Intel.*, 78:507-547.

Viviani, P. (1990). Eye movements in visual search: Cognitive, perceptual and motor control aspects. In E. Kowler (Eds.) *Eye Movements and Their Role in Visual and Cognitive Processes*, pp. 353-393. Amsterdam: Elsevier Science Publishers.

Williams, L. G. (1967). The effects of target specification on objects fixated during visual search. *Acta Psychol.*, 27:355-360.

Wolfe, J. M., Cave, K. R. and Franzel, S. (1989). Guided Search: an alternative to the feature integration model for visual search. *J. Exp. Psych. Hum. Percept. Perf.*, 15:419-433.

Yantis, S. (1996). Attentional capture in vision. In A. F. Kramer, M. G. H. Coles, and G. D. Logan (Eds.) *Converging Operations in the Study of Visual Selective Attention*, pp. 45-76. Washington, DC: American Psychological Association.

Yarbus, A. L. (1967). *Eye Movements and Vision*. New York, NY: Plenum Press.

Zelinsky, G. J. and Sheinberg, D. L. (1997). Eye movements during parallel-serial visual search. *J. Exp. Psych. Hum. Percept. Perf.*, 23:244-262.

Zipser, K., Lamme, V. A. F., and Schiller, P. H. (1996). Contextual modulation in primary visual cortex. *J. Neurosci.*, 16:7376-7389.

12

Vision and Action in Virtual Environments: Modern Psychophysics in Spatial Cognition Research

Heinrich H. Bülthoff
Hendrik A. H. C. van Veen

The classical psychophysical approach to human perception has been to study isolated aspects of perception using well-controlled and strongly simplified laboratory stimuli. This so-called cue reduction technique has successfully led to the identification of numerous perceptual mechanisms, and has in many cases guided the uncoverage of neural correlates (see chapters elsewhere in this volume). Its limitations, however, lie in the almost complete ignorance of the intimate relationship among action, perception, and the environment in which we live. Real world situations are so different from the stimuli used in classical psychophysics and the context in which they are presented that applying laboratory results to daily life situations often becomes impractical, if not impossible. At the Max-Planck-Institute for Biological Cybernetics in Tübingen, we pursue a behavioral approach to human action and perception that proves especially well-suited for studying more complex cognitive functions, such as object recognition and spatial cognition. The recent availability of high-fidelity "virtual reality" environments enables us to provide subjects a level of sensory realism and dynamic sensory feedback that approaches their experiences in the real world. At the same time, we can keep the ultimate control over all stimulus aspects that are required by the rules of psychophysics. In this chapter, we take a closer look at these developments in spatial cognition research and present results from several different experimental studies that we have conducted using this approach.

12.1   Introduction

In recent years, the study of spatial cognition experienced a strong technology push caused by major advancements in two very different areas: brain imaging and virtual reality. Indeed, those who manage to combine the potential of both technologies receive considerable attention (e.g., Epstein and Kanwisher, 1998;

Maguire et al., 1998a, 1998b). In this chapter, however, we focus exclusively on the role of virtual-reality technology in the evolution of the field. In subsequent sections, we identify the major motivations behind this development, and we provide illustrative examples taken from our own laboratory. Our primary goal here is to provide useful information for making proper and effective use of virtual reality in cognitive science.

So what exactly is happening? What do we mean when we say "virtual reality technology is pushing spatial cognition research ahead?" The answer probably lies in the very nature of virtual reality (VR): VR is a technique that strives to create the illusion of experiencing a physical environment without actually being there (the concept of verisimilitude has been mentioned in this context). The virtual environments (VEs) thus created provide the researcher with a new experimental platform in addition to the natural environment and the classical highly reduced and abstracted laboratory settings. We will see later on in this chapter why VEs have earned a place alongside those other options, and why their role in studying human spatial cognition is strongly increasing (also see Darken, Allard and Achille, 1998 and Péruch and Gaunet, 1998). Some of the major factors are illustrated in the following paragraph.

The classical psychophysical methods that are used to investigate perception are characterized by the use of well-controlled but, compared to the real world, strongly simplified laboratory stimuli such as dots, plaid patterns, or random-dot stereograms. These abstracted stimuli often bear little resemblance to those occurring in the real world, but are nevertheless very useful for identifying low-level perceptual mechanisms. The study of higher-level cognitive behaviors such as object recognition, visual scene analysis, and navigation requires a different methodology. At this level, the intimate and extensive relationship among action, perception, and the environment plays an important role (Gibson, 1966, 1979). To unravel the mechanisms working at this level, it is less important to understand perception itself than it is to investigate its role in guiding actions. Moreover, it is questionable whether one can study such higher-level mechanisms using the abstracted stimuli that are typically applied in psychophysics. In navigation, for instance, we repeatedly use landmarks to decide where we are and where we should go next. It is obvious that such a behavior strongly depends on the abundance and specific appearance of these landmarks. Thus, a systematic study of cognitive behaviors such as navigation ideally uses a methodology that supports the control and manipulation of arbitrary complex stimuli in a reproducible way, at the same time allowing for the recording of natural behavioral responses to these stimuli. The technology that enables us to develop such a methodology has become available only recently. Advancements in computer graphics and display technology have led to the emergence of a new area of computer science and engineering, called VR. As stated before, VR is essentially a technique that creates the illusion of experiencing a physical environment without actually being there. This is accomplished by intercepting the normal action/perception loop: the participant's actions are measured and used to update a computer representation of a virtual environment, which is then presented to the participant by means of visual and other displays (e.g., haptic, tactile, auditory). This technique allows us principally to manipulate all aspects of the sensory stimulation as well as the effects of the participant's actions on these sensory experiences. As such, it enables us to study fundamental questions in human cognition.

In the following sections, we describe several motivations behind the increased usage of VEs in spatial-cognition research. In Section 12.2 we discuss some issues from a biological cybernetics point of view. Subsequent sections deal with the technology that enables us to use VEs, discuss stimulus control in the context of VEs, point out the increased level of stimulus relevance that VEs offer compared to traditional laboratory methods, identify spatial cognition in VEs as a new, interesting field, and give examples from our own VE laboratory.

12.2   Biological Cybernetics

Biological cybernetics is a subfield of biology that studies the complete cycle of action and perception in organisms. More specifically, it studies how organisms acquire sensory information, how they process and store it, and how they finally retrieve this information again to generate behavior. Such behavior – like moving through the world – in its turn alters the sensory information available to the organism and in doing so closes the action/perception loop. Systematic research of this complex feedback system requires fine control over those elements of the action/perception cycle that lie outside the organism (i.e., the world with which the organism interacts0. Intercepting the feedback loop by manipulating the "world" parameters alters the way in which action can influence perception. Such open- and modified closed-loop experiments are common practice in neuroethology (Reichardt, 1973; Heisenberg and Wolf, 1984) and sensorimotor studies (Hengstenberg, 1993). Virtual-reality techniques are now for the first time enabling us to perform similar experiments in the domain of complex human behavior, such as navigation through unknown cities (Mallot et al., 1998) or manipulating virtual objects with a haptic simulator (Ernst et al., 1998; Ernst, Banks and Bülthoff, 2000).

Figure 12.1 shows a basic diagram of the action-perception cycle from a biological-cybernetics perspective. The diagram symbolizes the flow of information between organism and environment. It is strongly simplified to make it easier to concentrate on the important elements for this discussion. For instance, the homeostatic processes that take place within an organism are not included, nor do we attempt to differentiate between types of behavior that do or do not induce changes in the environment.[1] In Fig. 12.1 we illustrate how we think VEs can be utilized to study the action/perception cycle. Inside the natural environment (the "real" one, if you wish), a second environment is created by means of human/computer interfaces. A smaller or larger part of the user's behavior is monitored by input

---

[1]Please refer to Bülthoff, Foese-Mallot and Mallot (1997) and references therein for more details.

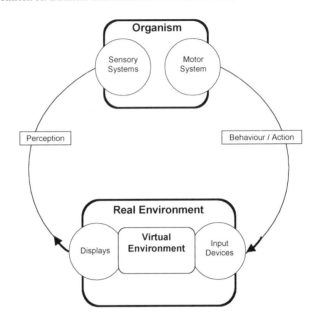

FIGURE 12.1. Information procession loop. The perception/action loop between sensory and motor systems is intercepted by a virtual environment that is encapsulated into the real environment. (Adapted from Fig. 9 in Distler et al., 1998a.)

devices such as movement trackers and is used to update a computer representation of the organism plus its virtual environment. Displays such as HMDs (helmet/head mounted displays) and earphones are used to communicate this representation to the user. Three important observations can be made by looking at this diagram. First, it is immediately clear that VEs are not substitutes for the real environment but are merely embedded in it. Thus, we must deal with a person experiencing two environments at the same time. Second, it is currently by no means possible – indeed it's hard to imagine being possible at all (but read Gibson, 1984) – to completely measure all behavior, to create a complete virtual world, and to stimulate all senses completely. A VE is always a reduced environment. Third, the devices interfacing the organism with the VE inherently suffer from delays, distortions, bandwidth restrictions, and limited ranges, which causes them to be distinguishable from the real environment. We will discuss these observations in more detail.

Two parallel worlds. VEs have been applied successfully for treatment of certain phobias (for an overview see Glantz et al., 1996, 1997), such as fear of spiders (Carlin, Hoffmann, and Weghorst, 1997), fear of height (Rothbaum et al., 1995), and fear of flying (Mühlberger et al., 2000). In the latter case, participants interact with a VE that simulates different stages of flying. Without ever leaving the ground, and with the participants fully aware of this and of the fact that everything is just a simulation, even a relatively simple VE can be convincing enough to induce fear and generate changes in physiological parameters such as heart rate, skin conductance, and EEG. Sometimes quite the contrary happens, and in a VE that has been designed for optimal visual quality participants do not feel immersed at all but rather start commenting on artifacts of the simulation such as that "all trees in the landscape look so similar." Sometimes it also seems as if participants can mentally switch from one world to the other and back, or even can observe both worlds in parallel! The central question here is: How much of the participants' perception and behavior is related to each of the worlds? What happens when the worlds provide conflicting information (such as in the aforementioned fear-of-flying treatment example)? Simple linear weighting models seem inappropriate here. In our eyes, the majority of the psychological and philosophical questions related to this concept of two parallel environments are yet unexplored. For some, this has been enough reason not to use VEs for spatial cognition research. A good way to see how much we can trust results obtained using VEs is to pair studies in VEs with studies in the natural environment. If the results gained in both environments are consistent with each other, further experiments can be performed in the VE taking advantage of its advanced features (see Section 12.4). We have done so, for example, in an experiment that studies mental representations of familiar environments. Inhabitants of the city of Tübingen in southern Germany were asked to point as accurately as possible toward well-known locations in their inner city, both while being present in the real city and while experiencing a very detailed virtual reality simulation of that same inner city (for details,see Sellen, 1998; Van Veen, Sellen, and Bülthoff, 1998). Subjects responded very accurately in both cases: the mean absolute pointing error was 11 deg when the subject was present in the real city, and increased marginally to 13 deg when experiencing the virtual version of that city. Further analysis showed that the pattern of systematic errors was extremely similar in the two conditions, suggesting that similar mental representations were recalled in both cases. Further experiments exploiting the simultaneous presence of real and virtual versions of this city environment are under way, as well as experiments in which we make changes to the virtual city that are not possible with the real one.

Incompleteness. Much can be said about the incompleteness of a VE in comparison to the real environment. If we focus on the direct implications for spatial cognition research using VEs, the most severe problem is probably the pitfall of superficial realism. A VE might look realistic enough for one's purposes but still can lack certain qualities that turn out to be essential for other tasks. For instance, after going to great lengths to create a realistic (mainly visually realistic) virtual model of the city of Tübingen (see Van Veen, et al., 1998; also see Fig. 12.2), at least one subject in our experiments (an inhabitant of real Tübingen) complained about the lack of appropriate height differences between the streets. She used to find her way around the town by remembering how certain roads sloped upward and others downward, something none of the other subjects seemed to do. Obviously, this is information of which researchers can make good use (work on the role of height differences in navigation is now under way, see Mochnatzki, Steck, and Mallot, 1999), but the potential danger is also clear. The system of validation elicited in the previous paragraph is again essential. Note, of course, that there are many other

FIGURE 12.2. Snapshot of virtual Tübingen. The 3-D reconstruction and rendering of a typical narrow street of historical Tübingen demonstrates the fidelity of our VR model which is achieved with few polygons but high-resolution texture maps for each individual house (there are not two houses in the 700-house model of Tübingen that are the same!).

obvious forms of incompleteness with which we also have to deal, such as the lack of stimulation of certain senses (typically only visual simulations are used in VR), the incompleteness of the stimulation (e.g., limited field of view), the simplicity of the environment, and all the problems associated with ego-movement in VEs. Some of these points are discussed again in Section 12.3.

Delays and distortions. An ideal interface between the participant and the VE should operate unnoticeably. If not, it's likely that participants start changing their behavior to circumvent the problems of the interface. Such change in behavior has, of course, implications for the validity of the experimental study. A typical problem is the feedback delay caused by the processing time required to reflect changes of the participants' behavior in changes on the displays. In vehicle simulators, for example, participants often compensate for feedback delays by reducing the speed of the vehicle (very slow speeds can mitigate the impact of feedback delays; see Cunningham and Tsou, 1999) and by employing alternate control strategies (Sheridan and Ferrel, 1963). Short delays are essential for studies involving fast control loops such as those found in steering tasks, manual manipulation, or head tracking. Distortions are especially evident and disturbing when parts of the real and virtual worlds interact, such as when the participant tries to grab a virtual object with his real hand or when head movements are measured to update the images displayed on the head-mounted display. While humans can adapt to delays and distortions (for reviews of spatial adaptation, see Harris, 1965, 1980; Welch,

1978; Bedford, 1993; for temporal adaptation, see Cunningham, Billock, and Tsou, 2000), this ability is limited (Bedford, 1999).

Two interesting concepts that are largely intertwined with the preceding discussion about the problems of the parallel worlds, incompleteness, and interfacing are presence and immersion. Slater and Wilbur (1997) define presence as

> *"a state of consciousness, the (psychological) sense of being in the virtual environment.",*

and immersion as

> *"a description of a technology, the extent to which the computer displays are capable of delivering an inclusive, extensive, surrounding, and vivid illusion of reality to the senses of a human participant."*

Their distinction between technology-related aspects and consciousness-related ones seems quite useful for better understanding why certain VEs are more effective than others.

12.3   Enabling Technologies

Given the way contemporary VEs are created, we can distinguish three different types of technologies: those that measure human behavior, those that support building virtual models, and those that display these environments to the user. We do not want to discuss these technologies here in full detail, but some key elements are worth mentioning, because they have helped to revolutionize our research.

Measuring behavior. The most interesting class of measuring devices with respect to spatial cognition research is the equipment that tracks the participant's movements through the real world. VEs are usually simulated within the confinement of a real room, and thus any type of ego-movement of the participant in the VE must be mapped onto movements within the boundary of that room. Often, the participant cannot move in the real world at all and must remain seated in front of a monitor or projection screen. Recent advancements in movement-tracking technology now allow for accurate real-time measurements of translation and rotation of the head, trunk, and hand within room-sized enclosures. In combination with a H-MD, the participant can move about in a virtual world by actually walking through a real space (e.g., see Chance et al., 1998; Usoh et al., 1999). The limited size of the real room remains a restrictive factor, of course. For studies involving larger virtual spaces different solutions are applied. In our laboratory, we use a specially configured exercise bicycle originally distributed by Tectrix™ and Cybergear™ (VRbike; see Fig. 12.3, and Distler, 1996) to move through large-scale virtual worlds such as cities and forests (see Distler et al., 1998). The participant needs to pedal and steer, the bicycle provides appropriate pedaling resistance and tilts in curves, but the whole configuration itself does not physically translate. We are therefore able to use this bicycle in front of a large panoramic projection screen.

FIGURE 12.3. VRbike in front of the large projection screen. The panoramic image of virtual Tübingen is projected by three ceiling mounted CRT projectors in such a way that at the head position of the cyclist a realistic 180 degree view of Tübingen can be experienced while cycling through the model. For more details see http://www.kyb.tuebingen.mpg.de/bu/projects/vrtueb

Similar solutions involve treadmills (e.g., see Darken et al., 1997) and car-like interfaces.

Building models. The requirements of other areas such as the military and game industries have led to the development of high-quality software and hardware for rapidly creating and rendering complex virtual environments. In our laboratory we make use of a very powerful graphical supercomputer (Onyx2^{TM} InfiniteReality^{TM}, manufactured by Silicon Graphics^{TM}) to reach a high level of visual realism. Note, however, that much cheaper PC-based systems are now also reaching performance levels that seem sufficient for many VE-studies of spatial cognition. At the software level, modeling tools such as 3D Studio Max^{TM} and Multigen^{TM} (which we use for many projects) offer tremendous capabilities for designing virtual worlds.

Display systems. Several different types of visual displays are in common use now. Simple monitors are used less and less due to the limited field of view that they provide and the restrictions on the participant's movements. In more recent years most of the technical developments have been focused on creating high-quality HMDs and panoramic projection systems. Truly panoramic systems are very expensive but can provide very high levels of immersion by covering the whole visual field with computer controlled imagery. HMDs are much cheaper and allow the subject to move around quite a bit more. Proper head tracking without delays is still extremely difficult though, and in practice HMDs often give disappointing results. HMDs do not cover the whole visual field with computer-generated images. Instead, their design effectively blocks sight of the real world in all directions and combine that with a small segment where the display is located. Other display types worth mentioning in relation to spatial cognition are auditory systems (for high-fidelity 3-D spatial audio rendering), haptic and tactile feedback systems (for providing contact cues with virtual objects), and motion platforms. These latter systems come in many varieties and are used to simulate physical movement of the observer, mainly by combining a little bit of real motion with a lot of transient motion cues (e.g., sudden onsets and offsets of motion, acceleration cues). The basic sensory systems that these devices stimulate are the proprioceptive and vestibular senses. We have installed a virtual reality system incorporating such a motion platform (manufactured by MotionBase^{TM}) in our laboratory in Tübingen, and it is currently being used for research on spatial updating and scene recognition. It will also be used to validate and extend the research on driving behavior that was done in our lab (Wallis, Chatziastros, and Bülthoff, 1997; Chatziastros, Wallis, and Bülthoff, 1998, 1999). Note that the VRbike mentioned arlier functions both as a measuring device (through its steering and pedaling sensors) and as a display (through its computer-controlled pedaling resistance and its tilting motion).

Much work is going on to improve all these technologies at many different levels. More immersive displays, more realistic environments, and more powerful motion trackers are under development and this will certainly improve the applicability of VEs for cognitive science.

12.4   Stimulus Control

Conducting experiments in VEs means that someone must program or define the complete content of the environment. Everything that is in there has explicitly been put there. This ensures that a precise description of the stimulus can be reported, allowing anyone to repeat or reproduce the experiment in order to validate the results. This is certainly not always possible with experiments in real environments. The major difference between conducting experiments in real and in virtual environments, however, is that in the latter case one has in principle complete control over the environment. This has several substantial advantages:

- all subjects can participate under exactly the same conditions

- the environment is optimally designed for the experiment

- no uncontrolled external factors in the environment (traffic, weather) can disturb the experiment

- any parameter of the experiment can be varied systematically, even during the experiment

- one can switch from environment A to environment B in a split second

- changes to the environment can be made at any time

FIGURE 12.4. Birds-eye view of a small city with a hexagonal street raster. This artificial city (Hexatown) surrounded by global landmarks served in several experiments to study the importance of local and global landmarks in human wayfinding.

We would like to demonstrate the power of extreme stimulus control by briefly summarizing a few experiments conducted in our laboratory.

Wayfinding and dynamic city layouts. In a series of experiments, Mallot and colleagues investigated the mental representation of spatial knowledge of structured large-scale environments (see Gillner and Mallot, 1998; Mallot, Gillner, Van Veen, and Bülthoff, 1998; Steck and Mallot, 2000). Using a specially created artificial virtual city called Hexatown (named after its hexagonal street raster, which forces a left–right movement decision at every junction; see Fig. 12.4), they tried to unravel the building blocks of mental spatial representation. To do so, subjects first learned certain routes through Hexatown until they could repeat them flawlessly. In the subsequent testing phase, subjects were instantly put at locations somewhere along the route and were then asked to start completing those routes. Between the training and testing phases, however, modifications to the city plan were made in such a way that different mental representations would correspond to different route completions. For instance, in Mallot and Gillner (2000) some of the buildings were moved to different locations. This way the researchers were able to conclude that the learned routes were stored in a graph-like representation of local elements, and not in a globally consistent survey map type of representation. Certainly no one outside Hollywood would consider conducting such experiments in the real world.

Visual homing in virtual worlds. Homing can be defined as the act of finding one's way back to a starting point after an excursion through the environment.

Communication with other organisms set aside, homing can be achieved by applying a combination of two basic mechanisms. In the "environment-centered" approach, the organism navigates by combining current-position information extracted from the local environment with its spatial long-term memory. In the "organism-centered" approach, the organism uses sensory information about its self-motion through the environment to continuously update its position relative to a starting point. Riecke and collaborators (Riecke, 1998; Van Veen, Riecke, and Bülthoff, 1999) studied whether this latter mechanism, usually called path integration, works properly and effectively when only visual information is present. They conducted triangle-completion experiments in high-fidelity vision-only virtual environments. On each trial subjects had to return to their starting point after moving outwards along two prescribed segments using the mouse buttons. Environment-centered strategies were precluded by replacing all landmarks in the scene by others during a brief dark interval just before the subjects started the return path. The results indicated that subjects acquired a fairly accurate mental representation of the triangular paths just by optical information alone. Omitting the scene modifications before the return movement resulted in nearly perfect performance, stressing the dominant role of environment-centered mechanisms under more natural conditions. Experiments such as this one obviously are extremely difficult to set up in the real world but can be done rather elegantly using VEs.

Scene perception and dynamic scene content. The process by which we recognize and analyze scenes remains largely mysterious. The evidence we do have suggests that the instantaneous, full, and detailed perception of a scene that we experience, is simply illusory and that detailed analysis of objects can only be achieved in a more piecewise, serial manner. In recent years, a phenomenon called change blindness has been used to estimate the accuracy of the representation of static scenes. Change blindness is the failure to detect a change in a scene, usually because the transient of the change is masked in some way (more can be found in other chapters in this volume). Wallis and Bülthoff (2000) have conducted an experiment in Tübingen in which they extended the change-blindness paradigm to dynamic scenes. A person drives or is being driven along a virtual road. At regular intervals the screen blanks for a very short period during which a change to the scene near the road is being made. Their results show that change blindness also occurs in dynamic scenes. In particular, they show especially that changes in object location are difficult to detect when the subject moves through the environment. Although others have managed to do related experiments in the real world (see Levin and Simons, 1997; Simons and Levin, 1998, and Chapter 10 by Simons and Mitroff in this volume), the level of systematic control available when using VEs is incomparable.

View-based scene recognition. Gibson (1979) showed the importance of the moving observer in a natural environment, but this importance extends to encoding and recognition of scenes also. If an observer knows where he is and in what direction he is looking, then by actively moving around he could build a coherent spatial representation of the immediate environment. The computer vision community has adopted the benefits of an active observer under the "active-vision paradigm", which is nicely illustrated in the book by Blake and Yuille (1992). Of course, psychologists have already known the importance of ego-motion and interactivity for a long time under the framework of perception for action. In a series of experiments, Christou and Bülthoff (2000) investigated how we represent our immediate environment. Specifically, they asked the question: If we learn to recognize a room from a limited set of directions, will we recognize it also from novel views? In the experiments, participants explored a virtual attic of a house (see Fig. 12.5) by using a six-degree-of freedom interface (Spacetec IMC Co., Massachusetts, USA) to drive a simulated camera through the environment. In the familiarization phase, participants had to find and acknowledge small encoded markers in the room that only appeared when viewed closely enough. The movement through the room was restricted along one major axis of the room and the viewing direction was restricted to the left or right by 60 degrees. Because the participants were only allowed to "walk" back and forth and could not turn around, they could never see the room from the other direction. After finding all the markers, each participant was shown pictures of the locations of each of the markers together with images from the other direction, which they never saw before. An equal number of distracter images taken from a similar 3-D "distractor" environment were also shown to participants. They simply had to respond when they believed the current image was taken from the original environment they had traversed during the familiarization stage. The results showed that after extensive, controlled, and yet realistic learning in a virtual environment, the restrictions imposed on the content of perceptual experience are still reflected in recognition performance. The familiar views were easily recognized whereas the performance dropped significantly for the novel views. Performance also dropped considerably when the active familiarization phase described previously was exchanged for passively watching a sequence of snapshots of the attic. The ability to recognize the novel direction views especially deteriorated. This was not the case for the back seat-driver condition, in which the active familiarization phase was replaced by passively watching a pre-recorded movie of another subject performing the active condition. In summary, Christou and Bülthoff have shown that active vision improves recognition performance. The back-seat driver condition shows that observer ego-motion is the critical variable, not volitional movement. What Christou and Bülthoff have not shown is what a more natural locomotion could provide. It is quite conceivable that recognition performance improves much more if observers are totally immersed in the virtual environment by either walking or cycling through it.

Stimulus control has been the key to the success of psychophysical studies of the past century. We hope to have shown that virtual-reality techniques now allow us to greatly extend the range of problems that can be studied with this approach.

FIGURE 12.5. Virtual attic. Experiments with active and passive exploration of this VR model helped us to understand the importance of the active observer in view-based scene recognition.

12.5   Stimulus Relevance

One of the hidden benefits of using VEs for spatial cognition research is the increased level of stimulus relevance. The classical reductionist's approach is to remove all stimulus components that are not directly relevant for the study being conducted. A single aspect of perception or behavior is singled out and studied in great detail, and all the other sensory inputs are kept to a minimum. Of course, we are all very much aware of the usefulness of this scientific method. Problems emerge, however, when we try to integrate the knowledge of all these isolated aspects to understand perception and behavior in natural environments. Nonlinear and dynamic interactions, a priori expectations (Bayesian vision!), interindividual differences, new levels of stimulus complexity, and highly dynamic scenes are only a few of the factors that often make such integration processes hopelessly complicated if not impossible. At the same time, one can ask whether the results obtained using isolated stimuli have any relevance at all for perception and behavior under natural conditions. Without claiming to have found a general solution to this problem, we would like to put forward the following consideration. In terms of perturbation theory, the classical reductionist's approach involves the systematic variation of one or a few stimulus parameters around certain control values, keeping all other parameters constant. The level at which all these other parameters are kept is often best described by "zero". However, perturbing the stimulus around "zero" is not a very ecologically interesting condition. Moreover, the reduced level of sensory stimulation might cause undesired changes in behavior that go unnoticed. A much more relevant approach, at least in terms of understanding perception and behavior in natural environments, would be to set all nonvaried stimulus aspects equal to a level typical of the natural environment. That obviously poses a stimulus-control problem because the number of parameters that would need to be considered is unimaginatively large. However, what we gain with such an approach is that we can assume that the perturbations in which we are interested are studied in a realistic context. In essence, we have greatly improved stimulus relevance. It is obvious that we want to conclude this consideration with expressing our belief that VEs can reach a level of sensory realism that is good enough to support such an approach. Whether or not this modern psychophysical method can live up to the promise of increased stimulus relevance remains to be proven, but for us it seems to be the only way out so far.

12.6   Spatial Cognition in VEs

An interesting development that spatial cognition researchers could employ is the increased use of VEs in different domains. Some people spend major parts of their working time in VEs, which gives spatial cognition in VEs a whole new meaning. Not only can we apply VEs for the study of spatial cognition but can study the spatial cognition of humans living in VEs. The problem itself is not completely new. For several decades simulators have been used to train driving and flying skills of military personnel, and gradually this approach has been transferred to the civilian domain. Obviously, the question of transfer of training from VEs to practical situations is related to the problem of validation that studies of spatial cognition using VEs must face. Advancements in technology and thinking are now creating new questions. What happens to the spatial mental representation of people confronted with temporally or spatially discontinuous VEs, created for instance by using hyperlinks (see Ruddle, 1999; Ruddle et al., 2000)? To what extent can we keep track of rapidly changing spatial scenes, such as those that can emerge when the historical development of a city area is (virtually) played back at high speed? What are the implications for spatial-information processing when the real world is augmented by overlaid spatial information generated from synchronous virtual models? Studying these and other unusual situations enabled by the new technologies might provide us with surprisingly new insights about the organization of our spatial memory and capabilities, especially with respect to plasticity and adaptability.

12.7   Concluding Remarks

We would like to point out here that we understand that VEs are not always the best way to go. Maximizing stimulus control is probably best achieved by removing all unnecessary cues from the stimulus; i.e., the classical reductionist's approach. Maximum stimulus relevance is of course only available in the natural environment. We hope to have made clear, however, that using VEs means combining the best of both approaches and opens up many new and exciting possibilities.

The introduction of VEs in spatial cognition research is along the same lines as the introduction of the gray-level raster display and the later extensive use of computer graphics in perception and recognition research. The increasing availability and falling costs of the technology will soon make these tools accessible to virtually anyone. We expect that within the early 2000s the use of VEs for studying spatial cognition will become common practice in many labs. The promise of increased stimulus control and relevance and the emergence of exciting new questions will certainly motivate many researchers to do so. Applying VEs will drive an integration process across disciplines: perception, behavior, and the (virtual) environment will reunite again.

In this light it might be worthwhile to briefly discuss the guest editorial called "Virtual Psychophysics" that appeared in the journal *Perception* (Koenderink, 1999). In his editorial, Koenderink first shows his excitement about the new possibilities that computers and virtual worlds seem to offer. He mentions several factors that are also highlighted in this chapter, such as increased stimulus control and stimulus relevance (which he considers enormously important), and he adds to that the benefit of being able to quickly produce all kinds of stimuli that "...would have been completely out of the scope of the old-day optical setups." But then he turns extremely skeptical and expresses his fear that most if not all of the modern psychophysical studies that use virtual worlds will turn out to be virtual psychophysics in a couple of decades. His main argument is that the visual realism of contemporary virtual environments is deceiving and almost nobody realizes that. He is really worried that "...present authors take familiarity with their virtual world pretty much for granted." or, in other words, that no one seems to care about a comprehensive description of the stimulus they use. We think that this view is way too skeptical. Of course, the apparent realism of contemporary VEs is only a trick, a trick that is getting better every year. The patterns of light and dark (the example used by Koenderink) shown on our displays are not the same as those encountered in the real world, even though they look pretty realistic to the untrained eye. But how important is that? And does not every researcher know that? Sure enough, for those of us who study how the distribution of light and dark in a scene conveys to us information about the detailed spatial relationships between scene elements, an extensive knowledge of the physical laws of optics and materials is essential. Indeed, for some of the problems in this specific area there is no piece of software that simulates the necessary level of physics. But we believe that a trained researcher will recognize such a situation and will refrain from using computer graphics in such a case. Similarly, those of us who study completely different problems, such as wayfinding, will judge the differences between the light patterns found in the virtual and real worlds as not or only marginally relevant to the task they are studying. In fact, they argue in much the same way as the reductionists argue when they remove every bit of stimulation that is not directly relevant to the task at hand, the only difference being the control point around which they conduct their perturbation studies. We believe that researchers are smart enough to realize that the computer graphics and virtual worlds they use are not the same as the real world. The patterns of light and dark are different, and so are the level of stimulus complexity, the level of sensory complexity, and the naturalness of movement among other things.

VE-based studies will be able to survive the test of time when we pay attention to two rules. First, those of us who want to generalize their results beyond the specific virtual world used for the experiment (which would otherwise indeed be nothing more than a study of spatial cognition in that particular VE), need to find ways to validate their results. This can be done by comparing the results with other studies, thus building up a framework of mutually supporting results, or by running similar experiments in the real world, which provides a framework itself. Second, and we support Koenderink in this, it is extremely important that scientists using VEs write down in their papers as complete as possible either how the particular virtual world (the stimulus!) has been created and displayed, or, alternatively, how it differs from the real world. With the expected developments in computer graphics in mind, this latter option might become more and more popular in the decades to come.

Acknowledgments

The authors would like to thank Stephan Braun for his help in preparing this chapter and Hartwig Distler for many stimulating discussions in the past that have helped to shape the insights presented in this chapter. While in Tübingen, Hendrik-Jan van Veen was funded by the Max-Planck Society and by the Deutsche Forschungsgemeinschaft (MA 1038/6-1, 1038/7-1).

References

Bedford, F. L. (1993). Perceptual learning. *Psychol. Learn. Motiv.*, 30: 1-60.

Bedford, F. L. (1999). Keeping perception accurate. *Trends Cog. Sci.*, 3:4-11.

Blake, A. and Yuille, A. L. (1992). *Active Vision*. Cambridge, MA: MIT Press.

Bülthoff, H. H., Foese-Mallot, B. M., and Mallot, H. A. (1997). Virtuelle Realität als Methode der modernen Hirnforschung (translation: Virtual reality as a method for modern brain research). In H. Krapp and T. Wägenbauer (Eds.), *Künstliche Paradiese – Virtuelle Realitäten*, pp. 241-260, Wilhelm Fink Verlag, München.

Carlin, A. S., Hoffmann, H. G., and Weghorst, S. (1997). Virtual reality and tactile augmentation in the treatment of spider phobia: a case report. *Behav. Res. Ther.*, 35:153-158.

Chance, S. S., Gaunet, F., Beall, A. C., and Loomis, J. M. (1998). Locomotion mode affects the updating of objects encountered during travel: The contribution of vestibular and proprioceptive inputs to path integration. *Presence: Teleop. and Virtual Environ.*, 7:168-178.

Chatziastros, A., Wallis, G. M., and Bülthoff, H. H. (1998). Lane changing without visual feedback? *Perception*, 27(suppl.): 59.

Chatziastros, A., Wallis, G. M., and Bülthoff, H. H. (1999). The effect of field of view and surface texture on driver steering performance (Utiliser un environnement virtuel pour évaluer indicateurs qui affectent la performance du conducteur). Proc. of Vision in Vehicles VII, Sept. 1997, Marseille, France. In A. G. Gale,, I. D Brown, C. M. Haslegrave, and S. P. Taylor (Eds.), *Vision in Vehicles VII*, Amsterdam: North-Holland/Elsevier Science B. V.

Christou, C. G. and Bülthoff, H. H. (2000). View dependency in scene recognition after active learning. *Mem. Cog.*, 27:996-1007.

Cunningham, D. W., Billock, V. A., and Tsou, B. H. (2000). Sensorimotor adaptation to violations of temporal contiguity and the perception of causality. Manuscript submitted for publication.

Cunningham, D. W. and Tsou, B. H. (1999). Sensorimotor adaptation to temporally displaced feedback. *Invest. Ophthal. Vis. Sci.*, 40:585.

Darken, R. P., Allard, T., and Achille, L. B. (1998). Spatial orientation and wayfinding in large-scale virtual spaces: An introduction. *Presence: Teleop. Virtual Environ.*, 7:101-107.

Darken, R. P., Cockayne, W. R., and Carmein, D. (1997). The omni-directional treadmill: a locomotion device for virtual worlds. Proc. UIST '97, October 14-17, 1997, Banff, Canada, pp. 213-221.

Distler, H. (1996). Psychophysical experiments in virtual environments. In *Virtual Reality World '96 Conference Documentation*, München: Computerwoche Verlag.

Distler, H. K., van Veen, H. A. H. C., Braun, S. J., and Bülthoff, H. H. (1998). Untersuchung komplexer Wahrnehmungs- und Verhaltensleistungen des Menschen in virtuellen Welten (The investigation of complex human perception and behavior in virtual worlds). In I. Rügge, B. Robben, E. Hornecker, and F. W. Bruns (Eds.), *Arbeiten und Begreifen: Neue Mensch-Maschine-Schnittstellen*, pp. 159-172, Münster: Lit Verlag.

Distler, H. K., van Veen, H. A. H. C., Braun, S. J., Heinz, W., Franz, M. O. and Bülthoff, H. H. (1998). Navigation in real and virtual environments: Judging orientation and distance in a large-scale landscape. In M. Göbel, J. Landauer, M. Wapler and U. Lang (Eds.), *Virtual Environments '98: Proceedings of the Eurographics Workshop in Stuttgart, Germany*, June 16-18, 1998, Springer Verlag, Wien.

Epstein, R. and Kanwisher, N. (1998). A cortical representation of the local visual environment. *Nature*, 392:598-601.

Ernst, M. O., Banks, M. S., and Bülthoff, H. H. (2000). Touch can change visual slant perception. *Nature Neurosci.*, 3:69-73.

Ernst, M. O., van Veen, H. A. H. C., Goodale, M. A., and Bülthoff, H. H. (1998). Grasping with conflicting visual and haptic information. *Invest. Ophthal. Vis. Sci.*, 39:624

Gibson, J. J. (1966). *The Senses Considered as Perceptual Systems.* Boston: Houghton Mifflin.

Gibson, J. J. (1979). *The Ecological Approach to Visual Perception.* Boston: Houghton Mifflin.

Gibson, W. (1984). *Neuromancer.* Great Britain: Victor Gollancz Ltd.

Gillner, S. and Mallot, H. A. (1998). Navigation and acquisition of spatial knowledge in a virtual maze. *J. Cog. Neurosci.*, 10:445-463.

Glantz, K., Durlach, N. I., Barnett, R. C., and Aviles, W. A. (1996). Virtual Reality (VR) for psychotherapy: From the physical to the social environment. *Psychotherapy*, 33:464-473.

Glantz, K., Durlach, N. I., Barnett, R. C., and Aviles, W. A. (1997). Virtual reality (VR) and psychotherapy: Opportunities and Challenges. *Presence: Teleop. Virtual Environ.*, 6:87-105.

Harris, C. S. (1965). Perceptual adaptation to inverted, reversed, and displaced vision. *Psych. Rev.*, 72:419-444.

Harris, C. S. (1980). Insight or out of sight? Two examples of perceptual plasticity in the human adult. In C. S. Harris (Ed.), *Visual Coding and Adaptability*, pp. 95-149, Hillsdale, NJ: Lawrence Erlbaum.

Heisenberg, M. and Wolf, R. (1984). *Vision in Drosophila.* Berlin: Springer Verlag.

Hengstenberg, R. (1993). Multisensory control in insect oculomotor systems. In F. A. Miles and J. Wallman (Eds.), *Visual Motion and its Role in the Stabilization of Gaze*, Elsevier Science.

Koenderink, J. J. (1999). Virtual psychophysics. Guest editorial, *Perception*, 28:669-674.

Levin, D. T. and Simons, D. J. (1997). Failure to detect changes to attended objects in motion pictures. *Psychon. Bull. Rev.*, 4:501-506.

Maguire, E. A., Burgess, N., Donnett, J. G., Frackowiak, R. S. J., Frith, C. D., and O'Keefe, J. (1998a). Knowing where and getting there: a human navigation network. *Science*, 280: 21-924.

Maguire, E. A., Frith, C. D., Burgess, N., Donnett, J. G., and O'Keefe, J. (1998b). Knowing where things are: Parahippocampal involvement in encoding object locations in virtual large-scale space. *J. Cog. Neurosci.*, 10:61-76.

Mallot, H. A. and Gillner, S. (2000). Route navigation without place recognition: What is recognized in recognition-triggered responses? *Percept.*, 29:43-55.

Mallot, H. A., Gillner, S., van Veen, H. A. H. C. and Bülthoff, H. H. (1998). Behavioral experiments in spatial cognition using virtual reality. In C. Freksa, C. Habel, and K. F. Wender (Eds.), *Spatial Cognition: An interdisciplinary approach to representing and processing spatial knowledge*, Lecture Notes in Artificial Intelligence Vol. 1404, Berlin: Springer Verlag.

Mochnatzki, H. F., Steck, S. D., and Mallot, H. A. (1999). Geographic slant as a source of information in maze navigation. In N. Elsner and U. Eysel (Eds.), *Göttingen Neurobiology Report 1999, Volume II, abstract No. 875*, Stuttgart: G. Thieme Verlag.

Mühlberger, A., Herrmann, M., Wiedemann, G., and Pauli, P. (2000). Treatment of fear of flying by exposure in virtual reality. Manuscript submitted for publication.

Péruch, P. and Gaunet, F. (1998). Virtual environments as a promising tool for investigating human spatial cognition. *Curr. Psych. Cog.*, 17:881-899.

Reichardt, W. (1973). Musterinduzierte Flugorientierung. Verhaltensversuche an der Fliege *Musca domestica. Naturwiss.*, 60:122-138.

Riecke, B. (1998). Untersuchung des menschlichen Navigationsverhalten anhand von Heimfindeexperimenten in virtuellen Umgebungen (Studying human navigation behavior by performing homing experiments in virtual environments). Unpublished Masters Thesis, Physics Department of the Eberhard-Karls-Universität Tübingen, Germany.

Rothbaum, B. O., Hodges, L. F., Kooper, R., Opdyke, D., Williford, J. S., and North, M. (1995). Effectiveness of computer-generated (virtual reality) graded exposure in the treatment of acrophobia. *Am. J. Psychiatry*, 152:626-628.

Ruddle, R. A. (1999). The problem of arriving in one place and finding that you're somewhere else. Proc. of the workshop on Spatial Cognition in Real and Virtual Environments, April 27-28, 1999, Tübingen, Germany, p. 58.

Ruddle, R. A., Howes, A., Payne, S. J. and Jones, D. M. (2000). The effects of hyperlinks on navigation in virtual environments, Manuscript submitted for publication.

Sellen, K. (1998). Schätzen von Richtungen in realen und virtuellen Umgebungen (Estimation of Directions in Real and Virtual Environments). Unpublished Masters Thesis, Biology Department of the Eberhard-Karls-Universität Tübingen, Germany.

Sheridan, T. B. and Ferrel, W. R., (1963). Remote manipulative control with transmission delay. *Percept. Motors Skills*, 20:1070-1072.

Simons, D. J. and Levin, D. T. (1998). Failure to detect changes to people in real-world interaction. *Psych. Bull. and Rev.*, 5: 644-649.

Slater, M. and Wilbur, S. (1997). A framework for immersive virtual environments (FIVE): Speculations on the role of presence in virtual environments. *Presence: Teleop. Virtual Environ.*, 6:603-616.

Steck, S. D. and Mallot, H. A. (2000). The role of global and local landmarks in virtual environment navigation. *Presence: Teleop. Virtual Environ.*, 9:69-83.

Usoh, M., Arthur, K., Whitton, M. C., Bastos, R., Steed, A., Slater, M. and Brooks, F. P. Jr. (1999). Walking > walking-in-place > flying, in virtual environments. Proc. ACM SIGGRAPH '99, pp. 359-364, Reading, MA: Addison-Wesley.

van Veen, H. A. H. C., Distler, H. K., Braun, S. J., and Bülthoff, H. H. (1998). Navigating through a virtual city: Using virtual reality technology to study human action and perception. *Future Generation Comp. Sys.*, 14:231-242.

van Veen, H. A. H. C., Sellen, K., and Bülthoff, H. H. (1998). Pointing to invisible landmarks in real and virtual environments. *Invest. Ophthal. Vis. Sci.*, 39:625.

van Veen, H. A. H. C., Riecke, B. E. and Bülthoff, H. H. (1999). Visual homing to a virtual home, *Invest. Ophthal. Vis. Sci.*, 40:798.

Wallis, G. and Bülthoff, H. H. (2000). What's scene and not seen: Influences of movement and task upon what we see. *Vis. Cog.*, 7.

Wallis, G. M., Chatziastros, A., and Bülthoff, H. H. (1997). Even experienced drivers have the wrong concept about how to change lanes. *Percept.*, 26(suppl.):100.

Welch, R. B. (1978). *Perceptual Modification: Adapting to Altered Sensory Environments*. New York, NY: Academic Press.

13

Selective Feature-Based Attention Directed to a Pair of Lines: Psychophysical Evidence and a Psychophysical Model

David Regan
Radha P. Kohly

The psychophysical data reported here indicate that observers can attend simultaneously to a pair of test lines while ignoring stimuli between the two test lines, and can extract independently four relationships between the two test lines. We measured discrimination thresholds for the mean orientation of a pair of lines as well as for their orientation difference, separation and mean location. We propose: (a) that all four discriminations were mediated by comparator mechanisms that received inputs from two first-stage narrow receptive fields whose centres were located some distance apart and which were "blind" to stimuli falling between the two receptive fields; (b) that the human visual system contains mechanisms of this kind whose outputs are labelled with the orientation difference, the mean orientation, the mean location, and the separation of the two first-stage receptive fields; (c) that orientation difference, mean orientation, separation, and mean locations are signalled independently. We found that all four discrimination thresholds were independent of test-line contrast for contrasts more than 2-3 times above line-detection contrast threshold. This finding can be understood if each of the four labelled outputs feeds an opponent-process stage. The preceding proposals can account for several previously reported phenomena. More generally, an array of the proposed long-distance comparator mechanisms constitutes a system that may be capable of fully specifying the shape, size, location and implicit orientation of the boundaries of an object's retinal image.

From 1815 to beyond the end of the century, the silhouette shown at the left in Fig. 13.1 would have been instantly recognisable in England. It can be argued that had the Duke been defeated at Waterloo, we would live in an alternative world in which French would be the first language for many more than in our present world – even in the alternative North America. But that aside, the first point to be made is that the outline in Fig. 13.1 is quite as easily recognisable as the solid silhouette – a point of considerable interest to the Gestaltists (Koffka, 1935; Ellis, 1967). Figure 13.2 shows that the same can be true even for a shape with a

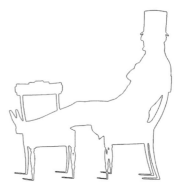

FIGURE 13.1. The Duke of Wellington. The solid silhouette and outline drawing are both instantly recognisable. Silhouette from E. Longford (1972), *Wellington: Pillar of State.* London: Weidenfeld & Nicolson.

complex internal structure. It is the everyday business of political cartoonists to provide remarkable examples of how a large amount of visual information can be transmitted by exploiting our exquisite sensitivity to the relative separation and orientation of lines. The photograph at the left of Fig. 13.3 is of Rowan Atkinson (a.k.a. Black Adder and Mr. Bean). Figure 13.3b is the familiar kind of poor likeness produced by an individual who draws incompetently. As illustrated in Fig. 13.3c, an individual with tolerable drawing skills can produce an excellent likeness with only a few strokes of the pen. A cartoonist can transcend this merely competent level of skill: by varying the relative location and orientation of a few penstrokes, the gifted cartoonist can exaggerate whatever it is that makes Rowan Atkinson look like Rowan Atkinson so as to create a drawing that is even more like Rowan Atkinson than Rowan Atkinson himself (Fig. 13.3). Indeed, in an earlier era, on seeing the man himself for the first time, people familiar with cartoons of Winston Churchill were taken aback, not only by his small size, but also by the somewhat unfamiliar appearance of the real person.

What is the basis in early visual processing for our exquisite ability to discriminate the shapes and configuration of line drawings and to effectively compare the shape of a solid and an outlined target? Our starting point in the present preliminary endeavour was a paper published by Morgan and Ward in 1985. These authors measured the just-noticeable difference in the separation between two test lines when the test lines were closely flanked by two additional lines whose locations varied from trial to trial in a way that was uncorrelated with the trial-to-trial variations of the separations of the test lines. Because the flanking lines were very close to the test lines, their variations of location would have corrupted the signals from first-stage spatial filters that "saw" both test lines. But the spatial jitter of the flanking lines did not affect discrimination thresholds for the separation between the test lines. Morgan and Ward concluded that, in their experiment, the line-separation discrimination threshold could not be explained in terms of the pattern of activity within first-stage spatial filters with strictly local receptive fields that "saw" both

FIGURE 13.2. The Duke of Wellington. The sketch of the Duke drawn from life is easily recognisable as is the outline drawing. Sketch from E. Longford (1972), Wellington: Pillar of State. London: Weidenfeld & Nicolson.

lines.

Morgan and Regan (1987) showed that the just-noticeable difference in the separation of two test lines was not affected by randomly varying the contrast of one of the lines on a trial-to-trial basis. Because this line-contrast manipulation caused random variations in the spatial Fourier transform of the test-line pair, this finding indicated that the visual processing that supported the discrimination took place in the spatial domain rather than the spatial-frequency domain. Morgan and Regan also reported that the discrimination threshold for line separation was independent of the contrast of the two lines, provided that contrast was more than 2-4 times above the line-detection threshold.

A proposed explanation for the several findings just described has been framed in terms of a long-distance interaction (Regan and Beverley, 1985, footnote 42; Morgan and Regan, 1987). It has been proposed that the human visual system contains a second-stage mechanism that receives input from two first-stage narrow receptive fields that are located some distance apart and has the following properties: (1) the mechanism is "blind" to any stimuli that fall between the two narrow receptive fields that feed it; (2) when both receptive fields receive optimal stimulation at the same instant, the output of the mechanism is much stronger than the sum of the responses to optimal stimulation of the two receptive fields one at a time. Thus, the hypothetical mechanism is a filter whose operation is essentially nonlinear. Morgan and Regan called their proposed mechanism a coincidence detector, although they reported no evidence that it could indeed respond to coincidences. Figure 13.4 depicts three such coincidence detectors.

Morgan and Regan (1987) found that the discrimination threshold for line separation was independent of contrast for line contrasts more than 2-3 times above the line-detection threshold. Their proposed explanation was that the discrimina-

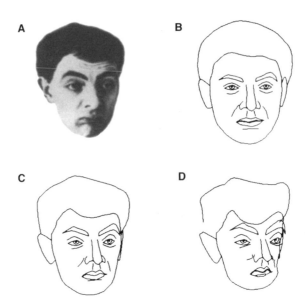

FIGURE 13.3. Rowan Atkinson. (a): Photograph. (b): A poor likeness. (c): A good likeness. (d): A cartoon. From G. Rhodes (1996). Superportraits. Hove, U.K: Psychology Press.

tion threshold for line separation is determined by the relative activity among a population of coincidence detectors that prefer different line separations, and they showed theoretically that an opponent-process version of their 'relative activity' proposal could account for their findings. The so-called combinatorial objection to this proposal is discussed in the Appendix to Kohly and Regan (2000).

To understand this proposal better we should step aside to discuss the finding illustrated in Fig. 13.5 that orientation discrimination threshold is also independent of contrast.

The proposal that orientation discrimination is based on opponency between orientation-tuned neurons can be understood by reference to Fig. 13.6. Two antennae are placed side-by-side, one of which (A1) broadcasts dots while the other (A2) broadcasts dashes. (It is essential that the beams overlap). A pilot flying exactly along the center of the overlap region hears a continuous tone, but a slight deviation to the left unbalances the signals and the pilot hears dots, whereas if the pilot deviates to the right he hears dashes. The angular accuracy of navigation can be very much higher than with one beam along, and much less than the angular width of either beam (Jones, 1978).

It will be useful to consider the factors that limit the smallest angular deviation $\Delta\theta$ that can be distinguished from noise (i.e., the angular threshold). In the first instance, let us suppose that, for either antenna, signal power S varies with direction θ according to $\cos^2 \theta$ (Fig. 13.6b). Clearly, the rate of change of S with respect to θ will be zero along the centre of the beam near the locally flat top of the $\cos^2 \theta$ lobe in Fig. 13.6b, but is positive for directions to the left of centre and negative

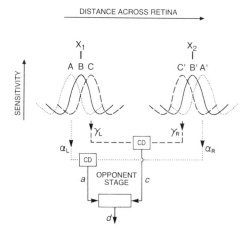

FIGURE 13.4. The coincidence detector of Morgan and Regan. Narrow spatial filters are connected in pairs to hypothetical coincidence detectors (CD). By definition, a coincidence detector does not respond to stimulation within the region between the two narrow filters that feed it. The output of this kind of a coincidence detector is labelled with the separation between its two input filters. The coincidence detectors feed an opponent-process stage whose output (d) signals line interval independently of line contrast and translation. From M. J. Morgan and D. Regan (1987). "Opponent model for line interval discrimination: interval and Vernier performance compared", *Vis. Res.*, 27:107-118.

for directions to the right of centre, being given by $\partial S / \partial \theta = -2 \cos \theta \sin \theta$ for a $\cos^2 \theta$ lobe. Depending on the shape of the lobe, the magnitude of $\partial S / \partial \theta$ may be maximal at some intermediate value of θ (as is the case for the Fig. 13.6b lobe shape), or may asymptote to a ceiling value, or may progressively increase as θ departs from zero.

If we ignore all factors other than $\partial S / \partial \theta$, directional discrimination will be best for the line of flight that gives the greatest difference in $\partial S / \partial \theta$ between the two beams. In the case of the Fig. 13.6b lobe shape, beam A1 should be directed to the left of the intended line of flight so that $\partial S / \partial \theta$ is maximal along the line of flight, and beam A2 should be similarly directed to the right of the line of flight. This idealised situation is illustrated in Fig. 13.6c.

There is, however, an additional factor. In all cases of interest to us, the signal power S declines monotonically as the magnitude of θ increases. In Fig. 13.6d, for example, where $\partial S / \partial \theta$ is shown progressively increasing with θ, there will be some point θ_M beyond which any potential reduction of angular discrimination threshold will be more than offset by the decline of power S. In Fig. 13.6d the region beyond this point is shown dotted. In practice, the optimal beam separation will be $2\theta_M$, as illustrated in Fig. 13.6d.

The third factor that limits directional discrimination is as follows. Even if the pilot flies exactly along the midline, faint dots or faint dashes will be heard from time to time because of irregular fluctuations in the relative signal strengths of the

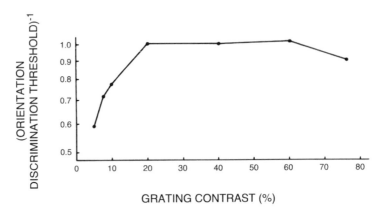

FIGURE 13.5. Effect of grating contrast on orientation discrimination threshold for a grating. Discrimination threshold is approximately independent of contrast for contrasts above 2-3 times contrast detection threshold. From D. Regan and K.I. Beverley (1985). "Post-adaptation orientation discrimination." *J. Opt. Soc. Am.* A2:147-155.

two beams. Errors in the direction of flight cannot be detected unless they produce dots or dashes that are appreciably louder than these random fluctuations in relative signal strength.

Noisy variations of receiver gain in the frequency range of the dots and dashes will also limit directional discrimination. Note, though, that directional-discrimination threshold is quite unaffected by fluctuations in signal strength that are common to both beams even when these fluctuations are large. Compare the situation when only one beam is used: large fluctuations of absolute signal strength produce a correspondingly large increase of directional discrimination threshold.

As an example of biological opponency, let us consider stimulus orientation. Suppose we have a population of cortical neurons, each of which is sensitive to orientation, spatial frequency, contrast, velocity, color, luminance, temporal frequency, and so on. We assume that the pattern of sensitivities is quite different for different neurons in the population. The following discussion parallels the discussion of Fig. 13.6 except that grating orientation or line orientation rather than direction is represented by θ, signal power S is replaced by neural sensitivity, and the two lobes, one of which is shown in Fig. 13.6b are neural-orientation-tuning curves.

We select a pair of neurons that prefer different orientations. A change of stimulus orientation will alter the relative activities of the two neurons. In this way, a more central neural mechanism that was sensitive to the relative activity of the two more peripheral neurons could discriminate a clockwise from an anticlockwise change of orientation that was much less than the orientation-tuning bandwidth of either neuron because, as illustrated in Fig. 13.6, orientation discrimination threshold would be limited by the difference in the slopes of the sensitivity curves of the two peripheral neurons, by uncorrelated noise in the two neurons' outputs

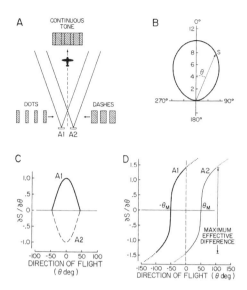

FIGURE 13.6. The Principle of Opponency. Modified from D. Regan (1986). "Form from motion parallax and form from luminance contrast: vernier discrimination." *Spatial Vision*, 1:305-318.

and possibly also by noise in the more central mechanism rather than by the orientation-tuning bandwidths of the two peripheral neurons. This property would be rather independent of the form of opponency. For example, it would hold for both subtractive and ratio opponency. Which pair of neurons would be most important for orientation discrimination can also be understood in terms of Fig. 13.6. This would be the pair whose $\partial S/\partial \theta$ differed most, allowing for the decline of signal strength with stimulus orientation θ (Regan and Beverley, 1985).

A sharp distinction can be drawn between the model depicted in Fig. 13.4 and models that are framed in terms of the outputs of filters that serve a single retinal location. For example, the line-element model of Wilson and Gelb (1984) represents a target as a point in a multidimensional spatial-filter space (see Wilson, 1991). This model is not equivalent to the model of Morgan and Regan (1987) because it is framed in terms of spatial filter outputs rather than in terms of the outputs of long-distance comparator mechanisms.[1]

---

[1] It is, of course, possible that a line-element model framed in terms of the outputs of coincidence detectors could account for the data of Morgan and Regan (1987), because line-element and opponent-process models share the common feature that they model discriminations that are based on the pattern of activity among a population of activated elements. However, this is not necessarily so. Although line-element and opponent models are fully equivalent for discriminations based on linear processes, they are generally not equivalent for discriminations based on nonlinear processing. Appendix D of Regan (2000) provides a comparison of the line-element and the opponent-process approaches to modelling discrimination data.

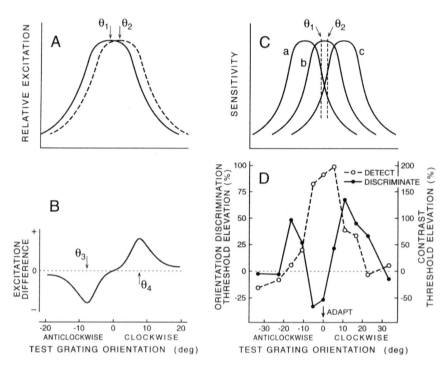

FIGURE 13.7. Evidence for opponent processing in the human visual system. (a) The notional excitation patterns produced within a population of orientation-tuned filters by gratings of orientations 1 (continuous line) and 2 (dashed line). (b) The difference between the two excitation patterns shown in panel A. (c) When the orientation of a stimulus grating changes from θ_1 to θ_2, the response of the most strongly excited neuron (b) changes negligibly, but there is a large change in the relative excitations of neurons a and b. (d) After adapting to a high- contrast vertical grating (zero on the abscissa) the contrast required to detect a test grating is elevated (dashed line and right ordinate). The detection threshold elevation is greatest for a vertical test grating. Orientation discrimination threshold is also elevated (continuous line and left ordinate), but the greatest elevation is for test gratings whose orientations are inclined at 10-20 deg from the vertical (θ_3 and θ_4). From D. Regan and K. I. Beverley (1985), "Postadaptation orientation discrimination," *J. Opt. Soc. Am. A2, 147-155.*

The output of the long-distance comparator mechanism (the so-called coincidence detector) defined earlier carries the label of line separation. In this chapter we report evidence that the human visual system contains long-distance comparator mechanisms that carry independent labels for the orientation difference, the mean orientation, and the mean location of a pair of lines.

What kind of psychophysical evidence can support a hypothesis that the human visual system contains a mechanism that is specialised for the discrimination of a given variable A? Before going further, we should note that there are several statements in the literature to the following effect. Suppose that the discrimination threshold for task-relevant variable A is measured in an experiment in which variable B is correlated with variable A. Suppose further that the discrimination threshold for B is measured in a second experiment in which B is the task-relevant variable. The claim is that a comparison discrimination threshold measured in the first experiment for variable A with the discrimination threshold for variable B measured in the second experiment provides a crucial basis, for the acceptance or rejection of the preceding hypothesis. We believe that this claim is incorrect. In particular, the finding that the discrimination threshold for task-relevant variable A is equal to or higher than the discrimination threshold for covarying variable B (measured separately with B the task-relevant variable) provides no solid evidence against the hypothesis that the visual system contains a specialised mechanism for discriminating variable A.

The relevant question is not, "is the discrimination threshold for task-relevant variable A lower than the discrimination threshold for variable B measured in a second experiment in which B is the task-relevant variable?" Rather, the relevant question is, "When A was the task-relevant variable, did observers base their discriminations on variable A and ignore task-irrelevant variable B?" We regard this as equivalent to the question, "Did the observer correctly perform the task?" We have developed a method for quantifying our confidence in the answer to the question just posed (Regan and Hamstra, 1992; Gray and Regan, 1997; Portfors and Regan, 1997). The advantage of the method is that it can be used in experiments where two or more variables covary within the stimulus set, a situation in which the partial regression coefficient procedure is unsatisfactory (Kohly and Regan, 1999).

13.1   Does the Visual System Contain Long-Distance Comparators with Orthogonal Orientation Difference and Mean-Orientation Labels?

The purpose of this experiment was to find whether the visual system contains long-distance comparator mechanisms whose outputs carry orthogonal orientation difference and mean orientation labels.

Fig. 13.8 illustrates the stimulus. There were two test lines and two "noise" lines. The noise lines were always placed between the test lines. To constrain the effective duration of the stimulus we presented a 100 msec masker immediately following each stimulus presentation. The width of the masker pattern was 1.5 times the maximum width of the four-line pattern. Optically superimposed on the monitor via a beam-splitting pellicle was a uniformly illuminated green screen which masked the slow phase of the phosphor afterglow of the line stimuli.

We varied α_T and β_T simultaneously and orthogonally to ensure that neither line alone provided a reliable cue to the task, that is, to force observers to base their responses on a comparison of the two test lines (see Fig. 13.8 for explanation of symbols).[2] We used a two-task design for the following reasons: (1) to allow us to test whether observers could discriminate α_T and β_T independently of one another, that is, to bring out the effect of changing the observer's task while keeping the stimulus set constant; (2) to allow a comparison of thresholds in the two-task and one-task conditions while keeping the stimulus set constant.

We randomly jittered the line length of the test and noise lines so as to remove trial-to-trial variations in the distance between the ends of the lines as a reliable cue to the task of discriminating orientation difference (see Regan et al., 1996).

There were six values of α_T, six values of β_T and six values of β_N, all symmetrically placed about zero. In most runs, the range of values was the same for α_T, β_T, and β_N (± 4 deg). Figure 13.9 brings out the point that within the set of 216 test stimuli there was zero correlation between α_T, β_T, and β_N (i.e. these three variables were orthogonal). All other variables were held constant: S_T was 49 arcmin and S_N was 7.5 arc min; M_T coincided with M_N.

Each trial consisted of one presentation of a test stimulus. The presentation duration was 20 msec. The observer's task was to signal after each trial whether (a) the test lines were turned inward (as in Fig. 13.8) or turned outward (i.e. whether $2\alpha_T$ was negative or positive), and (b) whether the mean orientation of the two lines was clockwise or anticlockwise of vertical.

Figures 13.10(a-f) show results for observer 1 with presentation duration 20 msec. It is evident from eyeball inspection that observer 1's responses were based on the task-relevant variable and that trial-to-trial variations of the task-irrelevant variables had little effect on the observer's responses. To express these points quantitatively, we measured the ratio (slope a)/(slope b) and the ratio (slope a)/(slope c).[3] These ratios were, respectively, 18:1 and 140:1. As in previous reports (Gray and Regan, 1999; Kohly and Regan, 1999), we defined the confidence ratio as equal to the smaller of the two ratios. The confidence ratio expresses our confidence that the observer' responses were based entirely on the task-relevant variable and ignored all task-irrelevant variables.[4] The corresponding threshold and confidence

---

[2]However, there is often a gulf between what should theoretically be true and what is true in practice when one uses electronic equipment, and especially where computers are concerned. As we describe later, we checked this point empirically. Some background to what might, to some, seem overcaution, "Clever Hans and worse", is narrated in Regan (2000, pp. 17-22).

[3]From this point on, by "the slope of the psychometric function" we mean the slope of the straight-line probit fit on probability paper.

[4]As discussed previously, although we cannot define some critical value of the confidence ratio above which we have 100% confidence, we consider that ratios above about 3 indicate a high level of confidence. But that even when a confidence ratio is close to unity we cannot conclude that the visual

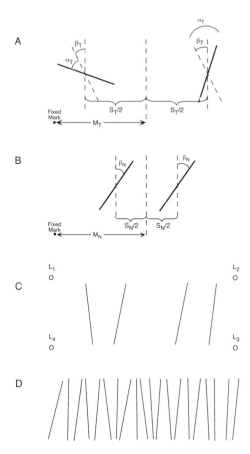

FIGURE 13.8. The stimulus. (a) The mean orientation of the two test lines was β_T deg, the difference between their orientations was $2\alpha_T$ deg, the separation between their midpoints was S_T deg of visual angle, and their midpoint was located M_T deg of visual angle from a fixed mark. (b) The mean orientation of the two "noise" lines was N deg, the separation between their midpoints was S_N deg of visual angle, and their midpoint was located M_N deg of visual angle from a fixed mark. The two pairs of lines were combined to create the stimulus illustrated in panel C. Note that the values of α_A, β_A, β_N, are considerably exaggerated in panels (a) and (b). Panel (c) gives a better impression of the values used in the experiment. L_1-L_4 in panel (c) were LEDs to aid fixation. Following each 20 (or 100) msec presentation a 20-line masker pattern centered on the location of the 4-line stimulus was presented for 100 msec. A typical masker patterns is illustrated in panel (d). From R. P. Kohly and D. Regan (2000). "Visual processing of a pair of lines: implications for shape discrimination." *Vis. Res.*, 40:2291-2306.

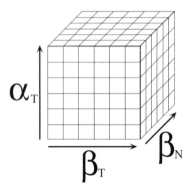

FIGURE 13.9. Stimulus Set Design. The set of 216 stimuli can be visualized as a $6 \times 6 \times 6$ cube of stimulus cells within which the three variables α_T, β_T, and β_N are orthogonal.

TABLE 13.1. Thresholds and confidence ratios for discriminating trial-to-trial variations in the mean orientations of the test lines obtained from the response data shown in Fig. 13.10.

Obs.	Discrim- ination Task	Presentation Duration (msec)	Orientation Difference Threshold (deg)	CR	Mean Orientation Threshold (deg)	CR
1	α_T and β_T	20	2.6 (0.1)	18	1.5 (0.1)	16
2	α_T and β_T	20	4.9 (0.2)	15	2.3 (0.1)	16
3	α_T and β_T	100	2.8 (0.3)	25	2.2 (0.2)	14

ratio for discriminating trial-to-trial variations in the mean orientation of the test lines were obtained similarly for the response data shown in Figs. 13.10d-f, and are listed in Table 13.1.

To rule out the possibility that some subtle stimulus artifact might have undone our rationale we carried out a control experiment in which only one test line was presented. Not unsurprisingly, the psychometric functions in Figs. 13.11c and f were flat, indicating that the observer ignored variations in the orientation of the noise lines. The ratio of the slopes in Figs. 13.11a and b was 0.90:1, not significantly different from unity. This implies that, when instructed to discriminate α_T, observer 1 was influenced by trial-to-trial variations in β_T to exactly the same extent as trial-to-trial variations in the task-relevant variable α_T. Similarly, the the slope corresponding to Figs. 13.10e and 13.10d was 1:1, and this was not significantly different from unity. Again, as expected, when instructed to discriminate α_T, observer 1 was influenced by trial-to-trial variations in β_T to exactly the same

system does not contain a specialised mechanism for the task-relevant variable (Kohly and Regan, 1999).

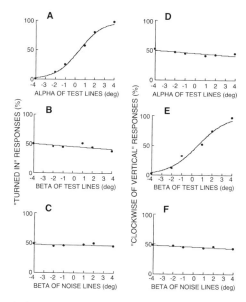

FIGURE 13.10. Discrimination of the difference of orientation)a-c) and the mean orientation (d-f) of a pair of test lines between which was a pair of noise lines. The task-relevant variable is plotted along the abscissas in (a) and (d). Task-irrelevant variables are plotted along the abscissas in (b), (c), (e), and (f).

extent as by trial-to-trial variations in the task-relevant variable β_T.

Spatial filters that "saw" both test lines must necessarily have 'seen' both noise lines also. We can reject the hypothesis that discrimination thresholds for orientation difference and mean orientation were based on the output of such filters on the grounds that the slope of the psychometric function in Fig. 13.10a is far steeper than the slope of the psychometric function in Fig. 13.10d and is similar to the slope of the psychometric function in Fig. 13.10e.

Now we turn to the question of orthogonality. Eyeball inspection of Figs. 13.10a, b, d, and e indicates that trial-to-trial variations of ± 4 deg in the mean orientation of the test lines had a negligible effect on the observer's responses when discriminating orientation differences, and that trial-to-trial variations of ± 8 deg in the orientation difference $(2\alpha_T)$ had a negligible effect on responses when discriminating mean orientation. We conclude that our observer could ignore mean orientation while discriminating orientation difference and could ignore orientation difference while discriminating mean orientation.

More quantitatively, changing the observer's task from discriminating orientation difference to discriminating mean orientation created the ratio slope(a)/slope (b) $\times$ slope(e)/slope (d) of 286. When the observer was denied a comparison between the two test lines (Fig. 13.11), this ratio fell to 0.96, a value not significantly different from unity.

How did observers 1 and 2 compare the two test lines? In principle, one way

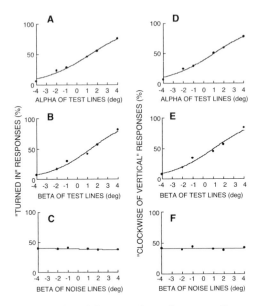

FIGURE 13.11. Same as Fig. 13.9 except that only one test line was presented.

would be to shift fixation from one line to the other. But this would not be possible within a total presentation duration of 20 msec; the shortest reported saccade latency is 100-150 msec (Kowler, 1990). A second way would be to shift the focus of attention from one line to the other while not moving the eyes. But the consensus seems to be that such a shift could not be achieved within 20 msec (Reeves and Sperling, 1986; Sperling and Weichselgartner, 1995).

Our findings can be explained along the lines proposed by Morgan and Regan (1987): long-distance comparator mechanisms that receive inputs from simultaneous stimulation of two narrow receptive fields some distance apart. We propose that discrimination thresholds were determined by such mechanisms whose outputs carried orientation difference and mean orientation labels and that these two variables were signalled orthogonally. These outputs were not affected by stimuli falling between the two narrow receptive fields that feed any given long-distance comparator mechanism.

13.2 Does the Visual System Contain Long Distance Comparators Whose Outputs Carry Orthogonal Mean-Location and Separation Labels?

We varied M_T and S_T simultaneously and orthogonally for reasons analogous to those set out in the previous experiment. We arranged that the maximum variation in M_T was exactly half of the maximum variation of S_T to ensure that the observer

TABLE 13.2. Discrimination thresholds and confidence ratios for each of the two tasks.

Obs.	Discrimination Task	Presentation Duration (msec)	Mean Location Threshold (arcmin)	CR	Separation Threshold (arcmin)	CR
1	S_T and M_T	20	6.2 (0.4)	4.7	5.2 (0.22)	16
2	S_T and M_T	20	5.2 (0.4)	4.9	8.6 (0.6)	16
3	S_T and M_T	100	4.0 (0.7)	3.9	5.6 (0.5)	14

could not unconfound the two variables by attending to one line only. As well, we randomly jittered the line length of the test and noise lines so that our observers could not use the aspect ratio of an imaginary rectangle as a cue to the line-separation discrimination task.

In the set of 252 test stimuli there were six values of M_T, six of M_N, six of S_T, and six of S_N. Test stimuli were divided into two subsets. In one subset of 216 stimuli, there was zero correlation between M_T, S_T, and M_N. In the second subset of 36 stimuli there was zero correlation between S_T and S_N. Within the first subset, the six values of S_N from the second subset were randomly allocated among the 216 stimuli and in the second subset the six values of M_T and M_N from the first subset were randomly allocated among the 36 stimuli. This design has been described previously (Kohly and Regan, 1999). Its purpose is to ensure that the observer cannot know from which subset any given test stimulus was drawn and therefore cannot adjust decision strategy according to subset. The range of values was 64 to 94 arcmin for S_T, 8 to 38 arcmin for S_N, and ± 7.5 arcmin for M_T and M_N. The values of S_T, M_T, S_N, and M_N chosen for the reference stimulus were equal to the corresponding means for the test stimulus set.

Each trial consisted of one 20 msec presentation of the reference stimulus followed by one 20 msec presentation of the test stimulus in what Macmillan and Creelman (1991, p. 135) call a reminder design. A 100 msec presentation of the masker pattern immediately followed both the reference and the test presentations. The two presentations were separated by a blank interval of duration 500 msec. The observer's task was to signal after each trial whether (a) the separation of the two test lines was greater or smaller than the separation of the two reference lines, and (b) whether the mean location of the test lines was to the left or right of the mean location of the reference lines.

Figures 13.12(a-f) show response data for observer 1 when presentation duration was 20 msec. It is evident from eyeball inspection that observer 1 based responses on the task-relevant variable and that trial-to-trial variations in the task-irrelevant variables had little effect on the observer's responses. To express this point quantitatively, we found the ratio of the slopes in a and b to be 7:1 and the ratio between the slopes in a and c to be 5:1. Discrimination thresholds and confidence ratios for each of the two tasks, obtained for our two observers, are listed in Table 13.2.

We carried out a control experiment in which only one test line was presented.

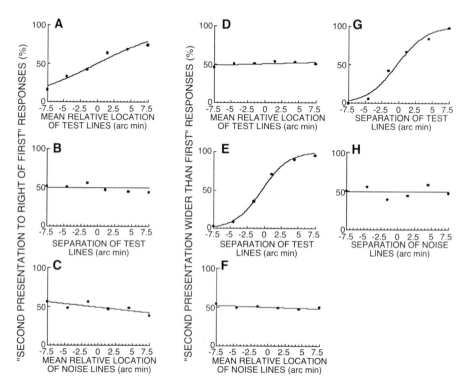

FIGURE 13.12. Discrimination of the mean location (a-c) and separation (d-f) of a pair of test lines between which was a pair of noise lines. The task-relevant variable is plotted along the abscissa in (a) and (d). Task-irrelevant variables are plotted along the abscissas in (b), (c), (e), and (f). From R. P. Kohly and D. Regan (2000). "Visual processing of a pair of lines: implications for shape discrimination," *Vis. Res.*, 40:2291-2306.

The pattern of results was quite different from that shown in Fig. 13.12. The psychometric function in the panels (a), (b), (d), and (e) of Fig. 13.3 were identical, indicating that the observer was totally unable to distinguish between trial-to-trial variations in M_T and S_T. In particular, the ratio of the slopes corresponding to Figs. 13.12(a) and 13.12(b) was 0.88, and this was not significantly different from unity. Similarly, the ratio between the slopes corresponding to Fig. 13.12(e) and 13.12(d) was 1.1:1, again not significantly different from unity. Not unsurprisingly, the observers ignored variations in both the separation and mean location of the noise lines.

Spatial filters that "saw" both test lines must necessarily have "seen" both noise lines also. We can reject the hypothesis that discrimination thresholds for separation and mean location were based on the outputs of such filters on the grounds that (1) that the slope of the psychometric function in Fig. 13.12(a) is 5 times steeper than the slope of the psychometric function in Fig. 13.12(c) and (2) the slope of the psychometric function in Fig. 13.12(e) is more than 100 times steeper than the slope of the psychometric function in Fig. 13.12(f).

When observer 1's task changed from discriminating the mean location of the two test lines to discriminating their mean separation, the ratio slope (a)/slope (b) × slope (e)/slope (d) was 265. When observer 1 was denied a comparison between the two test lines, this ratio fell from 250 to 0.99, a value not significantly different from unity. How did observers 1 and 2 compare the two test lines? As in experiment 1, we can reject a shift of fixation or a shift of attention because our presentation duration was 20 msec. Following the line of argument set out earlier, we conclude that our observers based their discriminations of M_T and S_T on the task-relevant variables and ignored all task-irrelevant variables. We propose that the two discrimination thresholds were determined by long-distance comparator mechanisms whose outputs carried mean location and separation labels, and that these two variables were signalled orthogonally.

13.3   Does the Visual System Contain Long-Distance Comparator Mechanisms Whose Outputs Carry Orthogonal Orientation Difference, Mean Orientation, Mean Location and Separation Labels?

We used a four-task design. Each trial consisted of a single 100 msec presentation of a test stimulus. Following each trial, the observer signalled whether: (a) the two test lines were turned inward or outward; (b) the mean orientation of the test lines was clockwise or anticlockwise of vertical; (c) whether test-line separation was larger or smaller than the mean of the stimulus set; (d) whether the mean location of the test lines was to the left or to the right of the mean location for the stimulus set. The set of 216 test stimuli comprised 6 subsets of 36 stimuli in each of which two variables had zero correlation. These were: α_T and β_T; α_T and S_T; α_T and M_T; β_T and S_T; β_T and M_T; S_T and M_T. Our observer could discriminate each of the task-relevent variables while ignoring all task-irrelevent variables.

13.4   How Do Discrimination Thresholds for Orientation Difference, Mean Orientation, Separation, and Relative Mean Location Vary as a Function of Contrast?

The stimulus sets and procedures were the same as in the first two experiments with the exception that observers performed a one-discrimination task rather than a two-discrimination task and that the contrast of the test lines varied across each threshold measurement.

Figures 13.13(a-d) show that all four discrimination thresholds were approximately independent of line contrast[5] for line contrasts greater than 2-3 times the line-detection threshold for both observers 1 and 2.

13.5   Attentional Implications and a Psychophysical Model

13.5.1   Long-distance comparators whose outputs signal orthogonally four stimulus attributes

Our *confidence ratio* procedure showed that observers ignored trial-to-trial variations in the noise lines and always based their discriminations on the task-relevant variables. We can therefore conclude that all mechanisms discussed here are indeed "blind" to stimuli located between the two test lines. We conclude that the human visual system contains coincidence detectors whose outputs are labelled with the mean orientation, orientation difference, and mean location, as well as the separation of a pair of test lines. Our four-task experiment indicated that orientation difference, mean orientation, separation, and mean location of the two test lines were processed independently of one another. We conclude that the outputs of our proposed long-distance comparator mechanisms carry orthogonal labels for those four variables.

---

[5]There are several definitions of the contrast of a single isolated target such as a bright line. Rather than Weber contrast $[(L_2 - L_1)/L_1]$ or Michelson contrast $(L_2 - L_1)/(L_1 + L_2)$ we chose to use the following definition: contrast is equal to $(L_2 - L_1)/L_2$, where L_2 is the luminance of the line and L_1 is the luminance of the surround. According to Burr et al. (1985) the rationale for this definition is that the visual response to a line is regulated by local gain control that occurs before spatial summation and before the detection stage.

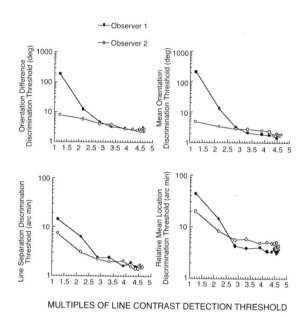

FIGURE 13.13. Effect of line contrast on four discrimination thresholds. effect of the contrast of the two test lines on discrimination thresholds for the orientation difference between the lines (a), their mean orientation (b), their separation (c), and their mean location (d).

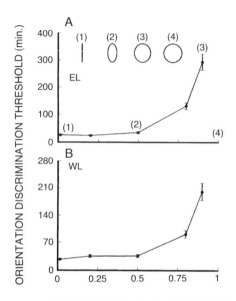

FIGURE 13.14. Implicit orientation I. Symbols show experimentally measured just-noticeable differences in the orientation of configurations ranging from a vertical line through vertically elongated ellipses to a circle (the just-noticeable difference is, of course, infinitely high for a perfect circle). The stimuli used to obtain various numbered points on the curve are illustrated with their corresponding numbers. Note that the orientation discrimination threshold for an ellipse with a 30 mm arc long axis and 15 arcmin short axis (inset 2) is approximately as low as for a 30 mm arc line (inset 1). From W. Li and G. Westheimer (1997). "Human discrimination of the implicit orientation of simple symmetrical patterns." *Vis. Res.*, 37:565-572.

It does not seem likely that attention-directed descending signals could switch visual-pathway connectivity during a 20 msec presentation. We conclude that visual processing takes place in parallel for mean orientation and orientation difference, mean location and separation, and orientation difference and separation. We suggest that our proposed long-distance mechanisms can respond to coincident stimulation of the two first-stage filters.

Following the algebra set out in the Appendix of Morgan and Regan (1987), we can understand why the four discrimination thresholds we measured were independent of line contrast for contrasts more than about 2-3 times above line-detection threshold, if we make the following additional assumptions. (1) Discrimination threshold for the mean location of two test lines was determined by opponent processing within a population of long-distance comparator mechanisms, each of which preferred a different mean location and whose outputs carried a mean-location label. (2) Discrimination threshold for the mean orientation of the two test lines was determined by opponent processing within a population of long-distance comparator mechanisms, each of which preferred a different mean orientation and whose outputs carried a mean-orientation label. (3) Discrimination threshold for orientation difference was determined by opponent processing within a population of long-distance comparator mechanisms, each of which preferred a different orientation difference and whose outputs carried an orientation-difference label.

13.5.2   Possible role of long-distance comparators in other psychophysical findings

Our proposal that the human visual system contains long-distance comparator mechanisms whose outputs orthogonally signal the mean orientation of the two test lines independently of their difference in orientation might account for the finding reported by Li and Westheimer (1997) that observers can discriminate the implicit orientation of a crossed pair of lines or the implicit orientation of an ellipse (Figs. 13.14 and 13.15).

Our finding that observers can dissociate and discriminate simultaneous trial-to-trial variations in the separation and in the difference in orientation of two lines provides independent evidence in support of the hypothesis put forward by Wilson and Richards (1989) that the curvature of a line is encoded in terms of the separation and difference in preferred orientation of two narrow spatial filters that are fed from distant locations (Fig. 13.16).

Our proposal that the human visual system contains long-distance comparator mechanisms that orthogonally signal both the separation of two lines and the difference in their orientations might account for our finding that observers can make acute discriminations of both Vee angle and the angle contained by crossed lines even when there are large random trial-to-trial rotations of the Vee or cross (Regan and Hamstra, 1992; Chen and Levi, 1996; Regan et al., 1996).

Long-distance comparator mechanisms that signal the mean location of the two test lines independently of their separation, orientation difference, and mean orientation would encode the local location of what has been termed the *core* of a shape (Burbeck and Pizer, 1995 and see Fig. 13.17).

Finally, the proposed mechanism whose output carries a line separation label could account for the finding that the aspect-ratio aftereffect caused by inspecting a solid sharp-edged rectangle transfers to an outlined ellipse (Regan and Hamstra, 1992). It could also explain why we can recognise a given shape whether it is the shape of a solid or the shape of an outlined figure - a problem of historical interests to the gestaltists (Koffka, 1935; Ellis, 1967).

13.5.3   Possible role of long-distance comparators in everyday vision

What might be the role of the mechanism we propose here in everyday human vision? We suggest here that it might detect objects and process the spatial attributes of their retinal images. The eye explores the environment by successively foveating

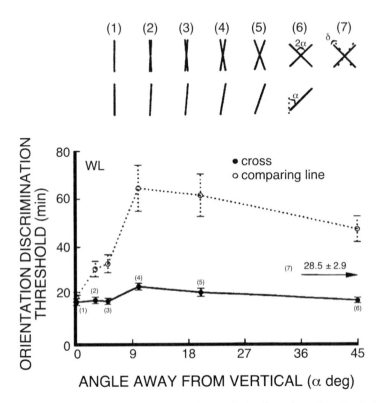

FIGURE 13.15. Implicit Orientation II. Solid symbols plot orientation discrimination thresholds for the axis of symmetry of a cross. The mean orientation of the axis of symmetry was vertical. The angle of the cross progressively increased along the abscissa, as indicated by the upper row of inserts. Progression was quantified in terms of the angle (α) between either line and the vertical The solid arrow indicates the orientation discrimination for a cross whose containing angle was randomly varied between 70 and 90 deg from trial-to-trial. Open symbols plot orientation discrimination thresholds for one of the lines that comprised the cross. From W. Li and G. Westheimer (1997). Human discrimination of the implicit orientation of simple symmetrical patterns. *Vis. Res.*, 37, 565-572.

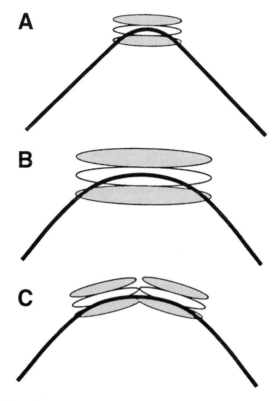

FIGURE 13.16. Schematic of curvature processing by orientation-selective spatial filters. Receptive fields are shown with an excitatory centre (white) and inhibitory flanks (gray). Although not shown, filters centred on the same locations but with other orientations are also involved in processing curvature. (a) A filter with a small receptive field tuned to high spatial frequencies would give a differential response to sharp curvature. (b) Filters with larger receptive fields would be required to cover the same extent of a more gradual curve. (c) An alternative scheme is to compare the orientations of two distant segments of the contour. From H.R. Wilson and W.A. Richards (1989). "Mechanisms of contour curvature discrimination." *J. Opt. Soc. Am. A*, 6:106-115.

different locations, resting for perhaps 200 msec on each location, and moving from one location to the next by executing a rapid saccade. When the opposite edges of the retinal image of an object fall on a pair of narrow receptive fields at the same time and the separation of the receptive fields is matched to the width of the retinal image, then the mechanism fed by the pair of receptive fields will be strongly activated. The same will hold for all the long-distance comparator mechanisms fed from opposite edges around the entire boundary of the object's retinal image, so that the system of mechanisms will signal the separation, mean location, mean orientation, and orientation difference of all the segments of boundary around the object's retinal image. As the eye saccades around the visual environment, it will often be the case that a sharp edge falls on only one of the two receptive fields that feed a given comparator mechanism. The mechanism will respond weakly, if at all. But in a cluttered visual environment, it will also often be the case that the two receptive fields that feed any given long-distance comparator mechanism receives simultaneous stimulation from two contours that belong to different retinal images. Therefore, if a system of such mechanisms is to function as an object detector, the simultaneity constraint will not be sufficient. Perhaps the Gestaltist concept of closure would add sufficiency (Koffka, 1935; Ellis, 1967). In particular, our hypothesis is that the proposed long-distance comparator mechanisms operate more effectively when stimulated by a closed outline figure than by line segments as in our present experiment. This hypothesis is currently being tested.

13.5.4 Attentional implications

There is considerable literature on focal spatial attention. Some authors have suggested that visual attention can be regarded as a spotlight (Posner et al., 1980), while others have compared it to a zoom lens (Eriksen and St. James, 1986). Yet other authors have proposed the idea of feature-based or object-based attention (Treisman and Gelade, 1980; Roelfsema et al., 1998).

It does not seem likely that our observers directed attention simultaneously to two sharply-defined locations, because they were able to carry out the task even when the location of any given line could not be predicted on a trial-to-trial basis (Kohly and Regan, 2000). We propose that: (1) each of the six possible pairings of lines in our four-line display drove a different population of long-distance comparators and (2) the output of each long-distance comparator encoded four orthogonally labelled quantities: orientation difference, mean orientation, mean location, and separation. (The concept of "label" is discussed in Kohly and Regan, 2000). The separation-labelled signal component would indicate which of the six populations of long-distance comparators were driven by the outer pair of lines. By selectively attending to this indicator the observer would, in effect, be attending to a *feature* (i.e. "outermost") that defined the test lines. This is a quite different from attending to two locations. Our model of this special case of feature-based attention accounts for how observers could, at the same time orthogonally process four relationships between two test lines while ignoring stimuli between those two lines.

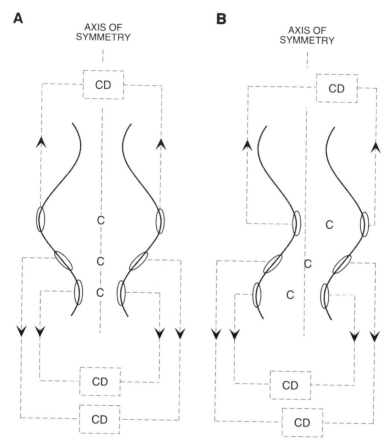

FIGURE 13.17. Extension of the coincidence detector concept. This hypothetical long-distance interaction mechanism differs from the one illustrated in Fig. 13.7 in that it is activated by simultaneous stimulation of two narrow receptive fields that are some distance apart and that may or may not prefer the same orientation. By definition, such a mechanism is not affected by stimuli in the region between the two narrow receptive fields. The output of the mechanism is labelled with the location (C) the point midway between the two receptive fields. From D. Regan (2000) *Human Perception of Objects: Early Visual Processing of Spatial Form Defined by Luminance, Color, Texture, Motion and Binocular Disparity*, Sunderland, MA: Sinauer.

Acknowledgments

We thank Dr. H. X. Hong for writing the display software. We thank Wei Hong for his patience as an observer and Eileen Kowler for advice. This research was supported by the Natural Sciences and Engineering Research Council of Canada (NSERC operating grant to D.R). D.R. holds the NSERC/CAE Industrial Research Chair in Vision and Aviation. Effort sponsored by the Air Force Office of Science Research, Air Force Material Command, USAF, under grant number F40620-97-1-0051. The US Government is authorized to reproduce and distribute reprints for governmental purposes notwithstanding any copyright violation thereon. The views and conclusions contained herein are those of the authors and should not be interpreted as necessarily representing the official policies or endorsements, either expressed or implied, of the Air Force Office of Scientific Research or of the US Government. R.P.K. was supported by an Ontario Graduate Student Scholarship in Science and Technology.

References

Burbeck, C. A. and Pizer, S. M. (1995). Object representation by cores: identifying and representing primitive spatial regions. *Vis. Res.*, 35:1917-1930.

Burr, D. C., Ross, J., and Morrone, M. C. (1985). Local regulation of luminance gain. *Vis. Res.*, 25:717-727.

Chen, S. and Levi, D. M. (1996). Angle judgements: Is the whole the sum of the parts? *Vis. Res.*, 36:1721-1735.

Ellis, W. D. (1967). *A Source Book of Gestalt Psychology*. New York, NY: Humanities Press.

Eriksen, C. W. and St. James, J. D. (1986). Visual attention within and around the field of focal attraction: a zoom lens model. *Percept. Psychophys.*, 40:225-240.

Gray, R. and Regan, D. (1997). Spatial frequency discrimination and detection characteristics for gratings defined by orientation texture. *Vis. Res.*, 38:2601-2617.

Jones, R. V. (1978). *Most Secret War*. London: Hamish Hamilton; reprinted 1998 by Wordsworth Editions Ltd., Ware, UK. (Published in the U.S.A. as *The Wizard War*. New York, NY: Coward, McCann and Geoghegan Inc.)

Koffka, K. (1935). *Principles of Gestalt Psychology*, New York, NY: Harcourt, Brace and World.

Kohly, R. P. and Regan, D. (1999). Evidence for a mechanism sensitive to the speed of cyclopean form. *Vis. Res.*, 39:1011-1024.

Kohly, R. P. and Regan, D. (2000). Visual processing of a pair of lines: implications for shape discrimination. *Vis. Res.*, 40:2291-2306.

Kowler, E. (1990). The role of visual and cognitive processes in the control of eye movements. In E. Kowler (Ed.) *Eye Movements and their Role in Visual and Cognitive Processes*. New York, NY: Elsevier, pp. 1-70.

Li, W. and Westheimer, G. (1997). Human discrimination of the implicit orientation of simple symmetrical patterns. *Vis. Res.*, 37:565-572.

Longford, E. (1972). *Wellington: Pillar of State*. London: Weidenfeld and Nicolson.

Macmillan, N. A. and Creelman, C. D. (1991). *Detection Theory: A User's Guide*. Cambridge: Cambridge University Press.

Morgan, M. J. and Regan, D. (1987). Opponent models for line interval discrimination: interval and vernier performance compared. *Vis. Res.*, 27:107-118.

Morgan, M. J. and Ward, R. M. (1985). Spatial and spatial-frequency primitives in spatial-interval discrimination. *J. Opt. Soc. Am. A.*, 2:1205-1210.

Portfors, C. V. and Regan, D. (1997). Just-noticeable difference in the speed of cyclopean motion in depth and the speed of cyclopean motion within a frontoparallel plane. *J. Exp. Psych.: Human Percep. Perf.*, 23:1074-1086.

Posner, M. I., Snyder, C. R. R., and Davidson, B. J. (1980). Attention and detection of signals. *J. Exp. Psychol. Gen.*, 109:160-174.

Reeves, A. and Sperling, G. (1986). Attention gating in short-term visual memory. *Psych. Rev.*, 93:180-206.

Regan, D. (1985). Storage of spatial-frequency information and spatial-frequency discrimination. *J. Opt. Soc. Am A.*, 2:619-621.

Regan, D. (1986). Form from motion parallax and form from luminance contrast: vernier discrimination. *Spat. Vis.*, 1:305-318.

Regan, D. (2000). *Human Perception of Objects: Early Visual Processing of Spatial Form Defined by Luminance, Color, Texture, Motion, and Binocular Disparity*. Sunderland, MA: Sinauer.

Regan, D. and Beverley, K. I. (1985). Postadaptation orientation discrimination. *J. Opt. Soc. Am. A*, 2:147-155.

Regan, D., Gray, R., and Hamstra, S. (1996). Evidence for a neural mechanism that encodes angles. *Vis. Res.*, 36:323-330.

Regan, D. and Hamstra, S. J. (1992). Shape discrimination and the judgement of perfect symmetry: dissociation of shape from size. *Vis. Res.*, 32:1655-1666.

Rhodes, G. (1996). *Superportraits*. Hove, UK: Psychology Press.

Roelfsema, P. R., Lamme, V. R., and Spekreijse, H. (1998). Object-based attention in the primary visual cortex of macaque monkey. *Nature*, 395:376-381.

Sperling, G. and Weichselgartner, E. (1995). Episodic theory of the dynamics of spatial attention. *Psych. Rev.*, 102:503-532.

Treisman, A. and Gelade, G. (1980). A feature-integrating theory of attention. *Cog. Psychol.*, 12:97-136.

Wilson, H. R. (1991). Psychophysical models of spatial vision and hyperacuity. In D. Regan (Ed.), *Spatial Vision*, pp. 4-86. London: Macmillan.

Wilson, H. R. and Gelb, D. . (1984). Modified line element theory for spatial frequency and width discrimination. *J. Opt. Soc. Am. A*, 2:124-131.

Wilson, H. R. and Richards, W. A. (1989). Mechanisms of contour curvature discrimination. *J. Opt. Soc. Am. A*, 6:106-115.

14

Thoughts on Change Blindness

J. Kevin O'Regan

That large changes in a scene are not noticed if they occur at the same time as a global visual disturbance caused by saccades, flicker, "mudsplashes", or film cuts, is generally explained in terms of a theory in which it is assumed that the observer's internal representation of the outside world is very sparse, containing only what the observer is currently processing. This chapter presents some clarifications of the theory, and some new implications and predictions that arise from it.

14.1   Introduction

A number of studies have shown that under certain circumstances, even very large changes in a picture can be made without observers noticing them (see reviews by Intraub, 1997; Simons and Levin, 1997; Simons, 2000). What characterizes all the experiments showing such blindness to scene changes is the fact that the scene changes are arranged to occur simultaneously with some kind of extraneous, brief disruption in visual continuity, such as the large retinal disturbance produced by an eye saccade (Ballard, Hayhoe and Whitehead, 1992; Grimes, 1996; Currie, McConkie, Carlson-Radvansky and Irwin, 1995; McConkie and Currie, 1996), a shift of the picture (Blackmore, Brelstaff, Nelson, and Troscianko, 1995), a brief flicker (Pashler, 1988; Phillips, 1974; Rensink, O'Regan, and Clark, 1995; 1997; 2000; Zelinsky, 1997; 1998), a "mudsplash" (O'Regan, Rensink and Clark, 1996; 1999), an eye blink (O'Regan, Deubel, Clark and Rensink, 1997; 2000), or a film cut in a motion picture sequence (Levin and Simons, 1997). Some examples of change blindness phenomena can be found on http://coglab.wjh.harvard.edu/ viscog/change and on http://nivea.psycho.univ-paris5.fr/

Current explanations (e.g., Rensink et al., 1997, 2000; Simons, 2000) of these phenomena are converging to a set of ideas concerning the nature of an observer's internal representation, the role of attention, and the role of visual transients, which might be summarized as follows:

1. The internal representation of the visual world is much sparser than subjective experience would seem to suggest.

2. Attention is required to encode an aspect of the scene into this representation.

3. Visual transients caused by scene changes can attract attention to the location of the change.

One example of where such ideas have been applied is in the work we have done on the flicker paradigm (Rensink et al., 1997, 2000), but similar views are implicit in the work using other paradigms. As suggested in this work, the reason why changes are often not seen when they are accompanied by flicker or other disruptions is that the disruptions create visual transients that swamp the local transient that would normally be created by the change. Because of this, attention is not attracted to the change location and observers must resort to other strategies to locate the change. Either they must search serially through locations of the scene in search of something that is changing or rely on their memory of the scene to determine what the change is. In both cases, changes will only tend to be noticed if they occur at locations that in themselves are likely to attract attention because they are somehow "interesting" to the observer.

The purpose of this chapter is to develop a number of points concerning this theory that have not been considered in detail before, and which lead to speculations and predictions for future work.

14.2   Thoughts on Normal Viewing: Where and What

This section addresses some questions concerning how the principles of the theory, as listed earlier, would apply to the situation in which observers *normally* see scene changes, that is, when no simultaneous, experimentally imposed disruption is superimposed at the moment of the change. I will begin by discussing some preliminary details concerning the notions of "central interest" and "marginal interest" which were employed in earlier papers. I will then come to the main point of the section, which concerns the distinction between the "where" component and the "what" component of the mechanism underlying normal change detection. This distinction will reoccur at different points in the rest of the chapter. It is important for methodological reasons and also gives rise to a counterintuitive prediction.

14.2.1   *Comments on the central/marginal interest distinction*

In work using flicker, blink, and mudsplash techniques, differences in the extent of change blindness were observed depending on whether the changing items corresponded to what was called "central" or "marginal" interest elements in the scene. The observed differences were important because they provided valuable arguments against the hypothesis that change blindness was caused by some kind of wiping-out of the internal representation provoked by the imposed scene disruptions – presumably a low-level wiping-out mechanism would incorrectly predict an equal wiping out of central- and marginal-interest items (see Rensink et al., 1997, 2000; O'Regan et al., 1999). Because of the importance of these arguments, it will be useful here to make the notion of "interest" more precise.

During examination of the scene, those particular aspects of the scene that will be attended to and encoded into the observer's internal representation will be determined by a variety of factors, ranging from low-level visual conspicuity (e.g. contrast, curvature, layout and color distribution in the picture), through semantic factors (e.g. coherence, relevance, prior knowledge), to individual preferences related to the observers' interests of the moment. Suppose that A, B, C,.... are aspects of a particular scene. A particular observer might attend to these aspects in the order: A, B, C, A, C, D, E, F, A, G, H,... A different observer, or the same observer on a different occasion, might attend to different items in a different order. But on average certain aspects of the scene will tend to be more often attended to and encoded than others, for example in this case A, B, and C. The notions of "central" and "marginal" interest (Rensink et al., 1997) were based on this idea: we defined "central interest" aspects as those aspects that an observer will be most likely to attend to and encode, and "marginal interest" aspects being those that an observer is least likely to attend to and encode.[1] We operationalized these definitions by asking a set of independent judges to verbally describe the pictures: we retained as "central interest" items those aspects that the judges always tended to mention, and as "marginal interest" items those aspects that the judges never mentioned.

A point that was perhaps not clearly made in our previous work was the fact that we considered this distinction between central and marginal interest to be merely a statistical, or operational way of defining aspects which observers will or will not generally attend to and encode. Theories of image comprehension might predict which items this might be, although factors of visual salience would additionally have to be taken into account. But for the purposes of understanding the phenomenon of change blindness, it is sufficient to assume that it is possible to classify scene aspects by order of likelihood of encoding.

14.2.2   Locations, objects, or aspects?

Another point about the central/marginal interest distinction is the following. It is tempting to assume that central- and marginal-interest aspects of the picture are *locations* in the picture. However evidence in the attention literature suggests that attention may perhaps better be described not in terms of space but in terms of objects: that is, not as attaching itself on the basis of spatially defined, contiguous regions in the visual field, but rather to (possibly noncontiguous) collections of attributes that can be grouped together to form objects (Duncan, 1984; Baylis, 1989; Baylis and Driver, 1993; Driver and Duncan and Nimmo-Smith, 1996). The notion of "object file" is also related to this (Kahneman, Treisman, and Gibbs,

---

[1] We are assuming that the observer encodes only what he or she sees. A possibility will be mentioned later according to which other aspects of the scene might be encoded *implicitly* without the observer being aware of them. However these implicitly encoded aspects would presumably be encoded into a different storage buffer, not available for making conscious comparisons, and changes in such aspects would therefore not be visible.

1992; Gordon and Irwin, 1996), as is Pylyshyn and Storm's (1988) notion of FINST. Despite the literature on this topic however, the term "object" seems unsatisfactory: Presumably an observer can, for example, attend to the sky (which is not really an object), or to a particular aspect of a scene (say, its symmetry or darkness) without encoding in detail all the attributes that compose it. Awaiting further clarification on this, it seems safer to suppose that attention can be directed to scene "aspects" rather than "objects."

14.2.3 Looking without seeing

An implication of looking without seeing is as follows. Looking directly at a scene location, an observer may be encoding a variety of aspects. For example, looking at an ambiguous figure, an observer may be encoding a young girl or an old woman. Looking at the middle of a word, an observer may be checking the typography of the middle letter of the word, or alternatively estimating the word's length, or else recognizing it. In each case the observer may be oblivious to the other aspects of the stimulus.

The idea that it should be possible to be looking at a region of the scene and only attend to a subset of the picture attributes that are there, is confirmed in an experiment we performed in which display changes were made simultaneously with observers' blinks (O'Regan, Deubel, Clark and Rensink, 2000). We found, by measuring eye movements while the observers searched for changes, that in almost 50% of the cases when an observer looked at a marginal interest location at the moment it changed, the change was not noticed, even though the eye was located less than 1 degree from the change.

Just as this is an example of "looking without seeing," there is equally the possibility of "seeing without looking." The eye may, for example, be fixating at the center of a circle, where there is nothing to be seen, in order to check that the circle is round rather than an ellipse.

14.3   The "Where" and "What" Components of Change Detection

Let us now return to the problem of seeing changes under normal circumstances and discuss how the points (1)–(3) underlying theories of change blindness would be used to explain what happens when in normal, everyday life an observer sees a change in a scene.

14.3.1 Transients tell "where"

Consider an observer contemplating a scene, and suppose that suddenly something changes. As is the case in many signal-processing devices, sudden changes in input generally provoke disturbances or "transients"[2] which can be detected by specialized spatiotemporal detectors (see Tolhurst, 1975; Breitmeyer and Ganz, 1976; Klein, Kingstone and Pontefract, 1992) that seem to have the special function of causing an "alert" that signals that a change has occurred (see Yantis and Jonides, 1990; Yantis, 1993), and of exogenously attracting attention to the location where the change was.

Note that the occurrence of the transient associated with the change has allowed the observer to detect *where* the change occurred. Turning attention to the change, the observer now sees the new scene element. However, unfortunately the observer no longer sees the *old* scene element. There may be some information contained in the nature of the transient that enables the observer to deduce what the change must have involved: the "flavor" of the transient may help in guessing, say, that an object has shifted rather than changed color. But the transient "flavor" itself will usually not be sufficient to determine exactly what the change consisted in. For this, the observer must have some memory of what was at the changed position before: that is, the observer must have at some earlier moment encoded into the internal representation what was previously at the change location.

14.3.2   Memory tells "what"

In summary, the role of the transient corresponding to the change is to provide information on *where* a change occurs, and the role of the internal representation is to allow the observer to know *what* the change was. This distinction between the "where" component and the "what" component of normal change detection is an important one and has not been sufficiently stressed in previous papers. One reason it is important is that it relates to the question of what is being measured in a change-blindness experiment: failure of the "where" component, failure of the "what" component, or failure of both? If change blindness experiments are used to estimate the richness of observers' internal representations of the visual world, care must be taken to ensure that it is the "what" component that is being probed: observers must really be required to indicate the nature of the change they have detected, and not merely the location of the change.

14.3.3   A counterintuitive prediction for change blindness in normal viewing

The distinction between the "where" and "what" components of change detection also leads to the following interesting consideration: previous work on change blindness has been restricted to experiments where some kind of global transient, like flicker etc., is presented simultaneously with the display change. But here we see that an implication of the theory is that, even when no such additional transients

---

[2]Analogously to electrical transients, that is, transitory large surges of current or voltage that occur in an electric circuit when it is turned on or off.

are added – that is, under completely normal viewing conditions – it should be the case that for what we have called marginal interest changes, though observers know *where* the change occurred, sometimes they should be unable to say exactly *what* the change was. This is because marginal interest aspects will tend not to be encoded. This prediction has not as yet been empirically tested. Indeed, a priori, the prediction seems extremely counterintuitive: we normally have the impression that any change in a scene, even in a marginal-interest aspect, is immediately visible.

An explanation of this incompatibility with common sense is perhaps first the fact that the prediction concerns marginal-interest aspects: because marginal-interest aspects of a scene are by definition not very significant to our conception of the scene, it may actually be the case that we are *not aware that we are not aware* of the exact nature of a change and satisfy ourselves with imprecise, "where" knowledge. Another, perhaps more likely, possibility is that normally changes in a scene occur slowly enough for attention or for the eye to reach them before they are completed. Finally, even when a change occurs too quickly for attention or the eye to reach it before it has completed, the "flavor" of the transient may often provide sufficient information about the change for the observer to feel satisfied he or she knows what it was.

On the other hand, cases should nonetheless exist when attention or the eye cannot quickly reach the change and where the "flavor" of the transient is not sufficient to guess what the change was. In these cases, observers, when interrogated, will have to admit that although they had the impression that something had changed in the scene, they cannot say exactly what the change was. This prediction, expected mainly for marginal-interest aspects of a scene, would be worth testing.

14.4   Thoughts on Disruptions

Having considered what happens in normal, everyday detection of changes in scenes, let us now turn to the situation generally studied in the literature on change blindness in which, at the moment of the change, some kind of additional, brief disruption of visual continuity is imposed by say, the concurrent occurrence of flicker, blinks, or eye saccades, for example. Rensink et al. (1997, 2000) proposed that such disruptions cause change blindness primarily because they interfere with the "where" component of change detection: they create additional transients in the visual field, which interfere with the attention-grabbing action of the local transient caused by the change. An observer's attention is no longer drawn immediately to the change location, and the observer must perhaps use a serial search strategy to locate the change.

14.4.1   Transients as masks and transients as distractors

But note that the interference with the "where" component of change detection can occur in two ways. First, the additional transients can act locally by combining with the transient corresponding to the change, rendering it less "attention-grabbing." This local combination can either involve luminance interactions or metacontrast-type masking effects, but in either case it makes the transient associated with the local change less salient and less effective in attracting attention to itself. Let us call this way of impeding the "where" component of change detection *local masking*.

A second way the brief disruption imposed by the experimental manipulation may interfere with the "where" component of change detection is by *diversion*. The local transient corresponding to the true change location is only one of a flood of additional, extraneous transients all over the scene. The "where" component of change detection is impeded because attention has no more reason to go to the "true" transient than to the many other, extraneous ones.

In most of the experiments performed in the 1990s on change blindness, both the local masking and the diversion mechanisms may be active in interfering with the "where" component of change detection. This applies, for example, to the eye saccade, flicker, blink and film cut experiments, which involve large disturbances in visual continuity all over the visual field. Although in the past we have given more attention to the diversion mechanism, it is clear that for a good understanding of the change blindness phenomenon, it would be worth studying in greater detail the separate influence of both the local masking and the diversion mechanisms. A start in this direction has been made with the use of the "mudsplash" and "masking rectangle" manipulations.

14.4.2   Measuring diversion with the mudsplash experiment

Instead of using a global disturbance at the moment of the change, in the "mud-splash" paradigm (O'Regan, et al., 1996, 1999) we briefly superimposed on the visual scene six small, high-contrast shapes, somewhat like mudsplashes on a car windshield. The positions of the mudsplashes were carefully chosen so that they did not cover the change location itself. In this experiment, therefore it was only through the mechanism of diversion that the imposed visual interruption interfered with the detection of the location of the transient associated with the change.

The results of the experiment confirmed that the diversion mechanism was sufficient to obtain a change blindness effect: a local masking of the transient associated with the change was not necessary.

This result is, of course, expected from the standard explanation of change blindness: the observer, not having a rich internal representation of the scene, attempts to rely on local transients to direct attention to the change location. But because there are now many rather than just a single local transient, the "where" mechanism of change detection is impeded and the observer's performance suffers.

14.4.3   The (critical?) number of diversions

An interesting question concerning the mudsplash experiment is the question of how many mudsplashes are necessary in order to prevent identification of the change. A first model would be to suppose that the effectiveness of the mudsplashes is determined merely by how many there are. The model would suppose that attention is attracted with equal probability to all the transients. The probability of moving to the transient corresponding to the true change would then just be $1/(N + 1)$, where N is the number of diverting mudsplashes. A better model might weight the probabilities by the relative salience of the transients, perhaps determined by their brightness, color, contrast, or other discriminating quality.

An alternative intuition one might have concerning the mudsplash paradigm would be related to the claim that there may be a critical number of events (say four or five) whose attention can be monitoring simultaneously (see Pylyshyn and Storm, 1988; Wolfe, Cave, and Franzel, 1989). It might therefore be that when fewer than four or five transients occur in the scene, verification of whether they correspond to a true scene change can occur easily, but when the number exceeds this critical value, change detection would suddenly break down. On the other hand, this view probably requires the notion of a visual buffer in which "what is currently being seen" is held. In that case, it would not be favored by the approach suggested here. Indeed, in pilot experiments presented in O'Regan (1998), I have found preliminary evidence against this view. Observers attempted to detect a letter that changed within an "alphabet soup" of scattered letters. The change occurred at the same time as a number of mudsplashes were spattered on the scene. The results showed that few large mudsplashes had an effect similar to many small mudsplashes, suggesting that the number of mudsplashes in itself is not the determining factor.

14.4.4   Proximity of the transient

An additional question concerning the diversion mechanism is the proximity of the diverting transients in relation to the true change position. It might be expected that a change would actually be more likely to be detected if an irrelevant transient occurs at a location quite near the true change location because now the transient is attracting rather than diverting attention from the area of the change. On the other hand, as we have seen from the idea that attention may not necessarily enclose all aspects of a region, and as suggested by the results of our "blink" experiment, one might equally well postulate the opposite, namely that spatial proximity would not be a direct determinant of probability of detection. This question would merit further investigation.

14.4.5   More questions on diversions

Many further interesting questions remain to be answered concerning how diversion can interfere with the "where" component of change detection. For example, can an observer learn to ignore diversions that have a known attribute or attend to transients that have a known attribute. If mudsplashes are always in the same locations, or if they are always of the same distinctive color, shape, or size, then an observer might be able to resist the tendency to orient attention to the diverting location. Similarly, if the local transient corresponding to the true change has some feature that makes it stand out within the flood of irrelevant local transients caused by the diversions, then change detection might become easier. On the other hand, it might be the case that the sudden onsets caused by transients cause an irrepressible attention-grabbing effect within which no selective mechanisms can operate. If the attention-grabbing action of the transients is an automatic, irrepressible, low-level mechanism, then it might be expected that the force of attraction might not depend on a computation determined by the other transients in the scene.

Another question concerns what kind of search strategy observers perform a-mong the locations that attract attention. Is there an inhibition of return mecha-nism that prevents attention from returning to a previously examined location? If so, how many previously visited locations can the inhibition of return mechanism keep in memory? Can this mechanism operate even though, unlike what happens in normal search tasks, the previously visited locations are repetitively creating attention-grabbing transients?

14.4.6   A transient pop-out task?

It must be stressed that although the mudsplash experiment used a purely diver-sional method to interfere with the "where" component of change detection, the measurements of change detection probability obtained from the experiment are still not a pure indication of the "where" component. The reason is that the "what," or encoding component of change detection presumably also played a role. De-pending upon whether observers construed the task they were accomplishing as consisting of merely indicating *where* the change occurred or as consisting of actu-ally determining exactly *what* the change involved, the role of observers' internal memory representations of the scene will be more or less great.

Indeed, probably a purer way of investigating how diversion contributes to change blindness would be to use a task where the "what," or encoding, com-ponent was not solicited at all. A possibility would be to use a kind of "pop-out" task in which observers must judge whether one of many transients provoked, for example by mudsplashes, appears different from the others and pops out.

14.4.7   Does local masking interfere with the "what" component?

The mudsplash experiment constituted an approach to the study of the diversion mechanism underlying the change-blindness effect. In the experiments in which there are global disturbances, such as eye saccades, flicker, blinks, and film cuts, there is also a local masking transient which, by combining with the transient at the change location, may decrease its salience. It would be interesting to study the relative weight, in interfering with the "where" component, of this local-masking transient compared to the diversion mechanism.

In particular, an important point concerns the possibility that the local masking transient actually somehow does more than just interfere with the "where" com-ponent of change detection. There is also the possibility that the local masking transient actually interferes with the "what," or encoding, component by wiping out the internal representation itself. For example, it could be postulated that the transient provokes a kind of "reset" of the internal representation in preparation for reception of new incoming information.

If this were true, then the basic tenet of our explanations of change blindness, namely that the internal representation is very sparse, could be discarded. We could claim that the internal representation is actually very rich but that it is wiped out by the local masking transient.

However this hypothesis has been addressed and rejected by several arguments (see Rensink et al., 1997, 2000; O'Regan et al., 1999). One of the main points is that it is incompatible with the finding that central- and marginal-interest scene aspects suffer different amounts of change blindness. A wiping out process would presumably wipe out both kinds of change equally. The mudsplash result, where there is no local masking at all, also argues against this claim.

14.4.8   Prediction for very slow changes

If we accept that the local masking effect imposed by flicker, for example, only acts on the "where" component of change detection, reducing the salience of the transient caused by the change, then it should be possible to manipulate the extent of this interference. One intriguing possibility would be if it were possible to arrange a situation in which aspects of a visual scene change so slowly that they do not generate visual transients at all. We would predict that changes created in this way would not be noticed if they were marginal- interest aspects, and pilot work we have done shows this to be the case.[3] D. Simons (personal communication) has also constructed film sequences where an object fades out without the observers noticing it.

This situation is perhaps similar to what happens when one looks at the hands of a watch: seeing a change in position can be done only by attributing a new code or classification to the current position (e.g., the minute hand was exactly on the 30-minute mark, but now has passed it). The same could be said of flowers: suddenly one is aware that they have wilted and need water, even though they are presumably slowly wilting all the time. In both cases, in order to "see" such a slow change, one must classify (encode) the new situation and judge it to be different from the previously encoded situation.

14.4.9   Estimating the "what" component of change detection using the masking rectangle experiment

Whereas the mudsplash experiment was (mainly) a way of studying the "where" component of change detection, the masking rectangle experiment (see O'Regan, Rensink and Clark, 1996, 1999) was a way of looking at the "what" or encoding component.

---

[3] Examples can be seen at http://nivea.psycho.univ-paris5.fr

Instead of diverting attention from the change location by means of extraneous mudsplashes, a black-and-white checkered rectangle was flashed over the area of the change while the change occurred. A large transient was therefore generated at the change location, presumably causing observers' attention to move to that region.

In this experiment, the location of the change is perfectly obvious, because it is indicated by the rectangle. The observers' task was thus clearly defined and was to say what was present before the masking rectangle appeared. Because the masking rectangle was of high contrast and completely covered the change location, the "flavor" of the transient provided no information about what was previously at the change, and subjects had to rely entirely on what information they had encoded prior to the appearance of the masking rectangle.

Coherent with the theory of sparse internal representation, we found that, particularly for the case of marginal interest changes, observers were often unable to report what the change was. Note that because in this experiment there are no diverting extraneous transients, the effects observed must be due wholly to the "what" (or encoding) component of the change-blindness phenomenon and not to any part caused by a "where"-related diversion component. The masking rectangle experiment therefore has the advantage over most other techniques used in change blindness research of representing a "pure" measure of the content of the internal representation.[4]

Note that in fact it would have also been possible to do the experiment without a masking rectangle at all by simply making the change directly. This corresponds to the situation discussed earlier where a scene change occurs in normal viewing without any additional visual disruption. However a disadvantage of using this technique in order to estimate the content of the internal representation is that because the transient caused by the change is not masked in any way, it will contain some information about the change that has taken place. Observers may make use of this "flavor" of the transient to deduce what the change might have been — for example a shift in an object may produce motion energy, which will be a different kind of transient from that produced by, say, a color change.

14.5   Other Issues Concerning the Theory

In the following sections, I will discuss some additional points about the explanation of change blindness based on the notion that the internal representation of the visual scene is sparse.

---

[4]Note that it must be assumed that the local masking caused by the masking rectangle does not wipe out the internal representation. The arguments against this were mentioned earlier.

14.5.1   A prediction for the moment of change detection

An interesting, so far unexplored consequence of this way of understanding change detection in the flicker, saccade, blink and film cut experiments concerns the time course of exploration of a scene. Let us recall what we assume happens when an observer examines a scene. Attention is directed sequentially to different aspects of the scene, and these aspects are encoded into a categorical, more durable memory store. What determines the order with which different scene aspects will be encoded is presumably low-level factors such as their visual salience, as well as high-level factors related to the process of object and scene identification. Suppose that the sequence of exploration for a particular scene involves attention being directed sequentially to aspects A, B, C, D, E, F, G, ..., labeled in descending order of "interest." The basic hypothesis of a sparse internal representation postulates that once a scene aspect A has been encoded, and once attention has moved onward to aspect B, the original scene aspect A is no longer being attended to and processed. If a change should now occur in A, and if conditions are arranged so that the visual transient created by the change is masked or camouflaged by one or another of the techniques currently being used in the change blindness literature, then the change should not be noticed. On the other hand, if the change had occurred while A was being encoded, the change would be noticed.

Because central interest aspects of a scene are presumably precisely those that will tend to be attended to at the beginning of picture contemplation, we can therefore make the curious prediction that even a change in a central-interest aspect of a scene may be missed if the change occurs after the scene has already been inspected for a while. In other words, we would expect that changes in the most significant aspects of a scene would be easier to detect *early* in the period of contemplation of the scene and harder to detect *later* in the period of contemplation. On the other hand, the opposite would be true of the less significant, marginal-interest aspects of a scene.

14.5.2   A prediction: seeing illusory appearances

A related prediction from this explanation of the scene-change experiments is that under some circumstances it should happen that an observer claims that a change has occurred in a picture when actually no change occurred at all. Suppose again that a picture contains aspects A, B, C, D, E, F, G, ..., in descending order of "interest." Suppose that before the transient the observer has encoded only the most "interesting" aspects, namely A, B, C, and D. Suppose that now a transient occurs in the region of element E, but that there is actually no change in the picture. It could now happen that on scanning the scene after the transient, the observer notices aspect E, which had not been previously encoded. The observer might well incorrectly deduce that this aspect had appeared during the transient. It is noteworthy that this kind of error might tend only to involve illusory appearances and not illusory disappearances.

14.5.3   A special role for layout?

Some studies in the 1990s have addressed the question of the particular role of layout in scene-change experiments (Simons, 1996; Simons and Wang, 1998). Under the conditions of these experiments, changes in layout are quite easy to detect, whereas changes in objects are more difficult to detect. This particular finding may perhaps be due to increased visual salience of the layout change. But additional findings, showing that detection of layout changes is not affected by verbal interference, whereas detection of object changes is, as well as other results showing a differential effect of changing the observer's viewpoint, all suggest that in fact layout plays a special role in perception.

On the other hand, one might make the following argument. It is clear that objects are never recognized in isolation. Modern theories of vision often suppose that visual analysis proceeds simultaneously at several spatial scales, so that information about an object always has associated with it information about the surroundings within which the object is situated. An analogy with music is appropriate: a melody played by a violin is perceived quite differently if the violin is playing as part of an orchestra.

If this is true, then the distinction between layout and object becomes less clear. In a task where the observer is looking at a scene consisting of objects A, B, C, D, E within a layout L, it could be that observers attending, say, first to object A and then to object C are in fact coding A+L, followed by C+L. Because the layout is always being encoded and put into memory storage, no matter what part of the scene the observer is processing (let us say it is element X) when a change subsequently occurs, then what is encoded after the change also involves the layout: X+L. If the layout has changed to L′, then the observer will encode X+L′ and the change in layout will always be noticed.

Despite this objection, it is nevertheless tempting to attribute a particular role to layout. Intuitively, it seems reasonable to suppose that the layout of a scene represents a sort of framework within which a picture is perceived. The situation may be similar to what is postulated in linguistics, where a distinction is sometimes made between "given" and "new" information. The "given" information is often the subject of a sentence, that is, what is known in advance and to be commented upon, and the new information or "topic" constitutes what is to be added to the "given" information (see Chafe, 1970; Haviland and Clark, 1974)

Another point is that there are reasons to believe that layout, because it constitutes an aspect of the environment which may be used in locomotion, navigation, and sensorimotor coordination, may be processed by different mechanisms, and may have a special status in determining perception (e.g. Milner and Goodale, 1995; Jeannerod, 1997). It is possible that the information-selection processes needed for processing layout may therefore be distinct from those that are used to make (verbal) decisions, judgments, and commentaries about objects. Indeed, if we say that what we mean by seeing is merely what is currently being processed, usually with a view to making decisions, judgments or commentaries about a visual stimulus, and if a separate, mainly motor-control-oriented process is usually dealing with layout, then why should layout changes be seen at all?

A possibility is that the separate, layout-processing mechanism, when it detects a change, creates some kind of attentional "alert" signal that is registered by the whole organism, including the processes that underlie what we usually call seeing. If this were so, then we would make the interesting prediction that the *fact* that a layout change has occurred might be registered, but not the exact nature of the layout change. Furthermore it might be expected that such an alert would only occur if the sensorimotor actions that would potentially be affected by the layout change were substantially altered. It might be possible to arrange situations in which two visually equally salient layout changes had different significance for sensorimotor coordination and thereby differed in detectability.

It is of course possible to arrange conditions where an observer must make (conscious) decisions and commentaries on scene layout, and in that case it must be the case that the observer really is "seeing" the layout. It would be interesting to see if the same layout change would be detected differently depending on whether the layout was being visually attended to (i.e. being "seen") or was being used implicitly in a sensorimotor task. It might be possible to arrange conditions where, when the observers are asked to visually attend to layout (i.e. "see" it), they do *less* well in detecting that a change has occurred than when layout is being made use of only implicitly in a sensorimotor task.

14.5.4   Implicit knowledge of changes?

An intriguing question about the scene change experiments is the question of whether, despite the fact that a change might not have been consciously noticed, it might nevertheless have been implicitly recorded so that it affects subsequent behavior.

One possibility concerns the low-level modules that analyse the incoming information in the first stages of visual processing. Some adaptation or modification of the functioning of these modules may occur through the mere presence of the information, and this may modify subsequent conscious or unconscious processing of the scene.

Another possibility relates to the fact that the visual system is not a unitary system. It is quite possible that, independently of the process that underlies conscious seeing, other processes (for example concerned with maintaining posture, adapting the grasp, or controlling eye movements) will have made use of the information (cf. Milner and Goodale, 1995; Jeannerod, 1997). Within these processes, some memory of the information, or at least some adaptation to its presence, may therefore have occurred, and this might affect the behavior of certain subsystems at a later time. For example, it is possible that eye movement scanning of the scene will be modified by the presence of the unseen elements. Some results (Hayhoe, Bensinger, and Ballard, 1998) are in support of this prediction, because they show that even though observers did not notice changes in blocks that they had to assemble in a block-copying task on a computer screen, their eye-fixation durations were nevertheless modified.

14.5.5   Other frameworks for explaining change blindness

In some recent papers (e.g. Rensink et al., 1997) an explanation of change blindness has sometimes been phrased in terms of the idea that attention is needed to see *changes*. Although this would be a possibility, in fact the theory that we have presented actually makes a more drastic claim, which is that attention is not only needed to see *changes*, but to see *anything at all*. Is there a real difference between the two views? As mentioned by Pashler (1995) it is not clear whether it actually makes sense to postulate a model of vision in which observers see everything in a scene, but, where, when a change occurs, they cannot see it unless they are attending to it. Such a view seems rather strained, and furthermore it runs the risk of espousing the "philosophically incorrect" position according to which there is an internal picture-like representation of the world that corresponds to what observers are currently seeing. It seems more theoretically coherent to take the extreme view that nothing is seen unless it is attended to. This is also the view taken by Mack and Rock (1998) in the context of their studies of "inattentional blindness".

Another question is raised by the proposition of Wolfe's (Wolfe 1997a, b; Wolfe, Klempen, and Dahlen, 2000) according to which change-blindness phenomena, and inattentional blindness, as well as phenomena that he has described using a "repeated visual search" paradigm, can all be understood in terms of the idea that everything is seen, but only what is attended to is remembered. Wolfe calls this "amnesic vision" or "inattentional amnesia." Again, as pointed out by Wolfe, Klempen, and Dahlen (2000), the distinction between the two approaches is largely a question of philosophical preference. Nevertheless Wolfe argues in favor of his view by claiming that, first, it better accounts for the subjective impression of visual presence that we have, and that, second, it accounts for the fact that in change blindness experiments, changes can be missed even when they are being directly attended to. However these arguments can be countered: first, as suggested in O'Regan (1992), the impression of complete visual presence that our subjective experience provides may be a sort of "solipsistic illusion" created by the fact that the slightest desire to see any part of the scene is immediately satisfied by a flick of the eye or of attention. The second counter argument to Wolfe is that in change-blindness experiments where apparently an attended change is missed, the observer, though looking directly at the change location as in Levin and Simons' (1997) film cuts and O'Regan et al.'s (2000) blink experiment, he or she might have been attending to an *aspect* of the scene at that location that was not the aspect that actually changed.

14.5.6   Relation to early literature on partial report

It is interesting to consider the relation between the more recent data on change blindness and the considerable literature that has accumulated on "iconic memory" since Sperling (1960) and Averbach and Coriell (1961) performed their classic experiments on partial report. Pashler (1998) in a review of this literature concludes that there is agreement about the fact that there probably exist two distinct forms of memory in the visual system: a high-capacity sensory memory which has a lifetime of the order of 100 msec, but which is sensitive to masking, and a low-capacity, more durable memory (usually called short-term memory), not sensitive to masking. A typical partial-report experiment involves extracting information from the sensory memory and transferring it into the more durable storage, and the observed phenomena can be adequately modeled by assuming that the choice of what is transferred is determined by attentional set, cues, and task instructions (Gegenfurtner and Sperling, 1993).

From such a point of view, what would be predicted concerning the present experiments? When a scene is presented to an observer, information from the scene impinges on the high-capacity sensory memory, and those parts of the scene that the observer wishes to process start being transferred into the more durable storage. After a while, this durable storage becomes full, and nothing more can be transferred. Now a change occurs in the scene. Because of the transients produced by the different experimental manipulations (e.g., saccades, flicker, mudsplashes), the observer has no cue indicating which part of the picture should be processed. Only if the observer, for whatever reason, (1) happens to decide to process the part of the scene that actually changed and (2) happened to have encoded that part of the scene into durable storage before the scene changed will a change be detected.

This is exactly the same analysis of change blindness as has been proposed in the change blindness literature. There is just one difference in the approaches, which is that in the iconic memory literature it is often implicitly supposed that the icon is what observers have the impression of seeing (this is similar to Wolfe's "amnesic vision" view), whereas here we prefer to suppose that observers only see what they are processing. Essentially, if we replace the notion of "more durable storage" or "short-term memory" with "what the observer is currently processing" and if we assume that the icon is not what is seen, then the two approaches are identical.

Pashler (1995) has also considered these questions and alluded to the fact that these distinctions may be too philosophical to be tested. However, he suggests a possible line of empirical investigation based on experiments in which he studied the effect of allowing the partial report stimulus array to be previewed prior to the appearance of the report cue (Pashler, 1984).

14.6   Conclusion

This chapter has examined in detail the implications of the four main assumptions made in current explanations of change blindness. Some distinctions that seem important for future work have been pinpointed, and some new experimental predictions have been made.

Perhaps the most important distinction that was made, and that permeated the reasoning throughout concerned not just the change-blindness situation but also the situation where in normal, everyday life a change occurs in the visual scene. The idea, which though quite clear and well-known had not previously been sufficiently stressed and developed, is that the process of detecting a change in the visual field involves two components: a "where" component and a "what" component. The "where" component provides information about the location of the change in the visual field, and is signaled by the local transient caused by the change. The "what" component allows the exact nature of a change to be ascertained and involves the use of information previously encoded in a durable memory store of some kind. The phenomenon of change blindness occurs because one or both of these components of normal change detection is interfered with.

Making precise the distinction between the "where" and "what" components of change detection led to the realization that even under normal viewing conditions – that is, when there are no experimentally imposed disruptions of the visual field – change blindness might still be found for certain marginal interest changes. Change blindness should also be found in the cases where changes are so slow that they do not create salient transients.

The where/what distinction also raised a methodological issue: To the extent that the precise task demanded of subjects in change blindness studies is sometimes not sufficiently clearly specified, these studies may not provide a pure measure of the content of the observer's internal representation. Future work should carefully control whether subjects are simply asked to locate a change, or whether they are also required to make precise judgments about the nature of the change.

Another series of points raised in this chapter concerned the precise mode of action of the experimentally imposed scene disruptions (e.g., eye saccades, flicker, blinks) in the change blindness experiments. These seem primarily to interfere with the "where" component of change detection, but do they do so by local masking or by diversion? Within the diversion mechanism, a number of questions for future work were raised.

Another consideration in the chapter was to make more precise the definition of "central"- and "marginal"-interest aspects of a scene that we had used in prior work. This led to a prediction about how change blindness might depend on the moment at which the change occurs, and to a prediction about the possibility of illusory appearances of changes.

Other points developed concerned the question of the role of layout, and the possibility of implicit perception of changes. Finally, some comments were made about other theories of change blindness and their relation to the early literature on partial report.

Acknowledgments

I thank R. Rensink, and J. Clark for teamwork during the experimental stages of this research, and Nissan Cambridge Basic Research for supporting it. I thank H. Ben Salah, S. Chokron, H. Deubel, V. Gautier, A. Gorea, T. Nazir, A. Noë, S. Shimojo, and J. Wolfe, and especially D. Simons, for their help and suggestions.

References

Averbach, E. and Coriell, A. S. (1961). Short-term memory in vision. *Bell Sys. Tech. J.*, 40:309-328.

Ballard, D. H., Hayhoe, M. M., and Whitehead, S. D. (1992). Hand-eye coordination during sequential tasks. *Phil. Tran. Roy. Soc. Lond. B*, 337:331-339.

Baylis, G. C. and Driver, J. (1993). Visual attention and objects: evidence for hierarchical coding of location. *J. Exp. Psych. Human Percept. Perf.*, 19:451-470.

Blackmore, S. J., Brelstaff, G., Nelson, K., and Troscianko, T. (1995). Is the richness of our visual world an illusion? Transsaccadic memory for complex scenes. *Percept.*, 24:1075-1081.

Breitmeyer, B. G. and Ganz, L. (1976). Implications of sustained and transient channels for theories of visual pattern masking, saccadic suppression, and information processing. *Psych. Rev.*, 83:1-36.

Chafe, W. (1970). *Meaning and the Structure of Language*. Chicago, IL: University of Chicago Press.

Currie, C., McConkie, G. W., Carlson-Radvansky, L. A., and Irwin, D. E. (1995) Maintaining visual stability across saccades: role of the saccade target object. Technical Report No. UIUC-BI-HPPP-95-01. Champaign: Beckman Institute, University of Illinois.

Driver, J., and Baylis, G. C. (1989). Movement and visual attention: the spotlight metaphor breaks down. *J. Exp. Psych. Hum. Percept. Perf.*, 15:448-456, 15:840 (1989) (erratum).

Duncan, J. (1984). Selective attention and the organization of visual information. *J. Exp. Psych. Gen.*, 113:501-517.

Duncan, J., and Nimmo-Smith, I. (1996). Objects and attributes in divided attention: surface and boundary systems. *Percept. Psychophys.*, 58:1076-1084.

Gegenfurtner, K. R., and Sperling, G. (1993). Information transfer in iconic memory experiments. *J. Exp. Psych. Hum. Percept. Perf.*, 19:845-866.

Gordon, R. D., and Irwin, D. E. (1996). What's in an object file? Evidence from priming studies. *Percept. Psychophys.*, 58:1260-1277.

Grimes, J. (1996). On the failure to detect changes in scenes across saccades, in K. Akins (Ed.) *Perception* Vancouver Studies in Cognitive Science Vol 2, pp. 89-110, Oxford: Oxford University Press.

Haviland, S. E., and Clark, H. H. (1974). What's new? Acquiring new information as a process in comprehension. *J. Verbal Learn. Verbal Behav.*, 13:512-21.

Hayhoe, M. M., Bensinger, D. G., and Ballard, D. H. (1998). Task constraints in visual working memory. *Vis. Res.*, 38:125-137.

Intraub, H. (1997). The representation of visual scenes. *Trends Cog. Sci.*, 1:217-221.

Jeannerod, M. (1997). *The Cognitive Neuroscience of Action*. Oxford, UK: Blackwell Publishers, Inc.

Kahneman, D., Treisman, A., and Gibbs, B. J. (1992). The reviewing of object files: object-specific integration of information. *Cog. Psych.*, 24:175-219.

Klein, R., Kingstone, A. and Pontefract, A. (1992). In K. Rayner (ed.) *Eye Movements and Visual Cognition: Scene Perception and Reading*, pp. 46-65, New York, NY: Springer.

Levin, D. T., and Simons, D. J. (1997). Failure to detect changes to attended objects in motion pictures. *Psych. Bull. Rev.*, 4:501-506.

Mack, A., and Rock, I. (1998). *Inattentional Blindness*. Cambridge, MA: The MIT Press.

McConkie, G. W., and Currie, C. B. (1996). Visual stability across saccades while viewing complex pictures. *J. Exp. Psych. Hum. Percept. Perform.*, 22:563-581.

Milner, A. D., and Goodale, M. A. (1995). *The Visual Brain in Action*. Oxford, UK: Oxford University Press.

Noë, A., Pessoa, L, and Thompson, E. (2000). Beyond the grand illusion hypothesis: What change blindness really teaches us about vision. *Vis. Cog.*, 7:93-106.

O'Regan, J. K. (1992). Solving the "real" mysteries of visual perception: The world as an outside memory. *Canadian J. Psych,* 46:461-488.

O'Regan, J. K. (1998). Detecting scene changes: an overview and a framework for recent findings. *Percept.*, 27(suppl.):36.

O'Regan, J. K., Deubel, H., Clark, J. J., and Rensink, R. A. (1997). Picture changes during blinks: Not seeing where you look and seeing where you don't look. *Invest. Ophthalmol. Vis. Sci.*, 38:S707.

O'Regan, J. K., Deubel, H., Clark, J. J., and Rensink, R. A. (2000). Picture changes during blinks: looking without seeing and seeing without looking. *Vis. Cog.*, 192-212.

O'Regan, J. K., Rensink, R. A., and Clark, J. J. (1996). Mudsplashes render picture changes invisible. *Invest.Ophthal. Vis. Sci.*. 37:S213.

O'Regan, J. K., Rensink, R. A. and Clark, J. J. (1999). Change blindness as a results of "mudsplashes". *Nature*, 398:34.

Pashler, H. (1984). Evidence against late selection: Stimulus quality effects in previewed displays. *J. Exp. Psych. Hum. Percept. Perf.*, 10:429-448.

Pashler, H. (1988). Familiarity and visual change detection. *Percept. Psychophys.*, 44:369-378.

Pashler, H. (1995). Attention and visual perception: Analyzing divided attention. In S. M. Kosslyn and D. N. Osherson (Eds.), *Visual cognition: An Invitation to Cognitive Science, Vol. 2* (2nd ed.), pp. 71-100. Cambridge, MA: MIT Press.

Pashler, H. (1998). *The Psychology of Attention*. Cambridge, MA: MIT Press.

Phillips, W. A. (1974). On the distinction between sensory storage and short-term visual memory. *Percept. Psychophys.*, 16:283-290.

Pylyshyn, Z. W. and Storm, R. W. (1988). Tracking multiple independent targets: Evidence for a parallel tracking mechanism. *Spat. Vis.*, 3:179-197.

Rensink, R., O'Regan, J. K., and Clark, J. J. (1995) Image flicker is as good as saccades in making large scene changes invisible. *Perception,* 24(suppl.):26-27.

Rensink, R. A., O'Regan, J. K., and Clark, J. J. (1997). To see or not to see: The need for attention to perceive changes in scenes. *Psych. Sci.*, 8:368-373.

Rensink, R. A., O'Regan, J. K., and Clark, J. J. (2000). On the failure to detect changes in scenes across brief interruptions. *Vis. Cog.*, 7:127-145.

Simons, D. J. (1996). In sight, out of mind: when object representations fail. *Psych. Sci.*, 7:301-305.

Simons, D. J. (2000). Current approaches to change blindness. *Vis. Cog.*, 7:1-16.

Simons, D. J. and Wang, R. F. (1998). Perceiving real-world viewpoint changes. *Psych. Sci.*, 9:315-320.

Simons, D. J. and Levin, D. T. (1997). Change Blindness. *Trends Cog. Sci.*, 1:261-267.

Sperling, G. (1960). The information available in brief visual presentations. *Psych. Mono.*, 74(11, Whole No. 498).

Tolhurst, D. J. (1975). Sustained and transient channels in human vision. *Vis. Res.*, 15:1151-1155.

Wolfe, J. M. (1997a). Visual experience: Less than you think, more than you know. In C. addei-Ferretti (Ed.), *Neuronal Basis and Psychological Aspects of Consciousness*. Singapore: World Scientific Press.

Wolfe, J. M. (1997b). Inattentional amnesia, In V. Coltheart (Ed.) *Fleeting Memories*. Cambridge, MA: MIT Press.

Wolfe, J. M., Klempen, N., and Dahlen K. (2000) Post-attentive vision. *J. Exp. Psych. Hum. Percept. and Perform.*, 26:693-716.

Wolfe, J. M., Cave, K. R., and Franzel, S. L. (1989). Guided search: An alternative to the feature integration model for visual search. *J. Exp. Psych. Hum. Percept. and Perform.*, 15:419-433.

Yantis, S. (1993). Stimulus-driven attentional capture and attentional control settings. *J. Exp. Psych. Human Percept. Perform.*, 19:676-681.

Yantis, S. and Jonides, J. (1990). Abrupt visual onsets and selective attention: Voluntary versus automatic allocation. *J. Exp. Psych. Human Percept. Perform.*, 16:121-134.

Zelinsky, G. J. (1997). Eye movements during a change detection search task. *Inv. Ophthal. Vis. Sci.* 38:S373.

Zelinsky, G. J. (1998) Detecting changes between scenes: a similarity-based theory using iconic representations. Beckman Institute for Advanced Science and Technology, Technical Report No. CNS-98-01.

Author Index

Subject Index